British Sociology's Lost Biological Roots

British Sociology's Lost Biological Roots

A History of Futures Past

Chris Renwick
University of York, UK

First published 2012 by
PALGRAVE MACMILLAN

Palgrave Macmillan in the UK is an imprint of Macmillan Publishers Limited,
registered in England, company number 785998, of Houndmills, Basingstoke,
Hampshire RG21 6XS.

Palgrave Macmillan in the US is a division of St Martin's Press LLC,
175 Fifth Avenue, New York, NY 10010.

Palgrave Macmillan is the global academic imprint of the above companies
and has companies and representatives throughout the world.

Palgrave® and Macmillan® are registered trademarks in the United States,
the United Kingdom, Europe and other countries.

ISBN 978-1-349-34737-7   ISBN 978-0-230-36710-4 (eBook)
DOI 10.1057/9780230367104

This book is printed on paper suitable for recycling and made from fully
managed and sustained forest sources. Logging, pulping and manufacturing
processes are expected to conform to the environmental regulations of the
country of origin.

A catalogue record for this book is available from the British Library.

A catalog record for this book is available from the Library of Congress.

10 9 8 7 6 5 4 3 2 1
21 20 19 18 17 16 15 14 13 12

For Clare

Contents

Figures

Preface and Acknowledgements

This book tells the story of how, in the first decade of the twentieth century, something happened that was particularly important but has been seldom discussed. After decades of debate, involving thinkers from a number of different fields, the British settled on an intellectual identity for sociology, the new general science of society. They did this by appointing L. T. Hobhouse, a man now better known for his liberal political philosophy, as the UK's first professor of sociology and the first editor of *The Sociological Review*, which was then the country's only sociology journal. This book argues that we have never fully appreciated the significance of that double appointment, in particular how very different British sociology could otherwise have been. The reason is that scholars have largely ignored one of the central concerns of those who participated in the debates leading to Hobhouse's appointments: how to relate sociology and biology. By recovering the options that were available in those discussions, this book explains how and why British sociology became a Hobhousean enterprise in which sociologists kept a critical distance from biology. In so doing, this book's aim is to change the way we talk about the origins of British sociology.

Because I am a practising historian, rather than a sociologist, this book's ambitions are framed in a very particular way. My primary aim is to faithfully reconstruct what happened during the founding British debates about sociology – a job I argue has not been done properly before – and, on that basis, offer an explanation as to why things turned out the way they did. Another, and no less important, aim, though, is to bring that history into a particular relationship with the present. As I explain at various points throughout the book, the history I tell is a distinctly intellectual history, in that the aim is to recover, as best we can, the intentions and motivations of those who wanted history to take a different course. In many ways, this is a perfectly respectable and conventional pursuit. However, my suggestion is that we make an explicit effort to bring that history to bear on the present, where biology is once again on the social science agenda, by considering what the past might do for sociology of the future. To be sure, I do not believe history should be 'presentist'. Nor do I think historians need to side with any of the actors in the histories they construct. Nevertheless, I am convinced that if history is about uncovering how the past has shaped the present then we need to consider the possibility that it can be a critical as well as documentary practice. In this sense, I hope to demonstrate that history can be understood as something that is full of potential, a pursuit that is as much of the future as it is of the past, rather than an enterprise devoted to the preservation of relics in the name of curiosity or entertainment.

These ideas were developed in a very particular intellectual environment, the world of history and philosophy of science, where I did my postgraduate training. Consequently, I am indebted to those who supported me during my MA and PhD studies in the Division of, now Centre for, History and Philosophy of Science at the University of Leeds, between 2004 and 2009. In addition to the Arts and Humanities Research Council, which funded both my postgraduate degrees, I owe my biggest debt to Gregory Radick, who supervised my research throughout those five years. As lead supervisor, Gregory encouraged me to pursue my research according to my own agenda, offered sound advice, and always delivered the goods when I needed him to. In this respect, I must also thank Graeme Gooday, my secondary supervisor, for his assistance, not least when I needed paid work to enable me to finish the thesis from which this book is mainly derived. Furthermore, I should recognize the input of everyone else at Leeds during those years, including Berris Charnley, Pete Vickers, Mike Finn, and Efram Sera-Shriar, who turned up to hear me present my research in seminars and then discuss it. Particular thanks here go to Jonathon Hodge, Richard Gunn, and Christopher Kenny who constantly offered their intellectual counsel during that process. As those who spent time with me at Leeds will know, this book is critical of certain historiographic developments that many of my contemporaries at Leeds support. However, I regard this book as a product of the Leeds HPS tradition and I hope everyone there will see it in that way too.

Looking further afield, I must register my appreciation for the efforts of those who helped set me on the path that led me to this point, in particular Paolo Palladino, Stephen Pumfrey, and Peter Harman, as well as the staff at the numerous universities and institutions I have visited along the way. Moreover, I wish to thank Philippa Grand at Palgrave Macmillan for taking an interest in my work and James Griffin for the book's cover design and all those who assisted with or provided feedback on earlier drafts, including Andrew Campbell, Thomas Dixon, Adrian Wilson, Mark Francis, Maggie Studholme, Mary Morgan, Harro Maas, Christopher Green, Bernard Lightman, Jon Agar, and referees for the *Journal of the History of the Behavioral Sciences*, *Isis*, and the *British Journal for the History of Science*, where some of the research from this book was first published. Special thanks here must go to Steve Fuller who has been a constant source of inspiration and encouragement. More than anyone else, Steve has always understood what I try to do in my work and, in the best spirit of academic debate, has been committed to helping other people understand it too.

However, since I wrote most of this book, I have moved from Leeds to York, where I have taken up a lectureship in the history department. I am incredibly grateful to the department for deciding my CV was worth looking at during what can be described only as a dispiriting round of job applications. Whilst the outgoing head of department, Bill Shiels, has made me feel welcome in my new institutional surroundings, Alex Goodall, Tom Pickles, Katherine Wilson, Catriona Kennedy, Hannah Grieg, Liz Buettner, Mark Roodhouse, Mark Jenner, Nick Guyatt, and Stuart Carroll have all helped me settle in to life at York. My thanks also go to Roger Burrows, Amanda Rees, Nick Gane and Mike Savage for not only making me feel welcome in the sociology department but also their efforts to forge the kinds of relationship between history and sociology that I talk about in this book.

Finally, I must acknowledge and give thanks to the people who have helped me in a personal but no less important sense. My parents, Shelagh and Stephen, have both worked hard and made countless personal sacrifices to ensure that I would have the opportunities, including the chance to attend university, which they did not. I hope this book is evidence I made the most of those opportunities and appreciate everything they have done for me. Most importantly, though, I must thank my long-suffering wife, Clare, for her love and support before, during, and after the writing of this book. Whilst Clare has tolerated my frequent work related absences, she has also shown great, and probably misplaced, faith in my ability to deliver on the promises I frequently make. Clare has always given me much more than she has received in return and hopefully this book is the first step towards making sure that things will be very different in the future.

Foreword

This is history of science at its best – a search for a moment of decision in the past that set in motion a train of events that constitute a now taken-for-granted way of organizing knowledge. However, this is not a search for origins in order to legitimise the present. On the contrary, Chris Renwick's aim is to reveal the openness of the original situation, especially since we are entering a period – a century after the one recounted in these pages – when a similar decision may be facing us. In the broadest terms, the decision is to do with negotiating the difference between the social and natural sciences – more specifically, the disciplinary boundary between sociology and biology. At stake is nothing less is what it means to be 'human', especially in a sense that requires a special body of knowledge somewhat set apart from the study of living things more generally.

Whatever decision one takes about where and how to draw the line between sociology and biology, some decision must (and had to) be taken. In the early years of the 21st century, as in the early years of the 20th century, there are strong intellectual and political currents calling for the social sciences to be fully subsumed under the natural sciences. In both cases, Darwin's name is talismanic. In terms of the candidates for the first UK sociology chair, which provides the centrepiece for Renwick's book, the person I have in mind here is Patrick Geddes, not Francis Galton. We nowadays think of Geddes mainly as a visionary urban and regional planner whose designs remain in evidence throughout the world from Mumbai to Tel Aviv. However, his inclination to treat human beings as animal populations adapted (or not) to their environments anticipate recent concerns that bring together evolutionary psychology, ecology, human geography, and the sociology of space and mobility. These movements are amongst those today that would welcome a more porous boundary between the natural and social sciences, even at the cost of ontologically levelling the distinction between the human and non-human. But my guess is that as the current century wears on and advances in biotechnology throw open new possibilities for intervention in the human condition, Geddes will prove not nearly as interesting as Galton, the founder of eugenics, whose own anthropocentric orientation cannot be denied.

Although the main benefactors of the London School of Economics (LSE), Sidney and Beatrice Webb, were clearly sympathetic to eugenics, a *de facto* chair in the field – held by Galton's follower Karl Pearson – had already been established at the University of London's flagship college. This bit of institutional politics probably did more to make Galton or another eugenicist a non-starter for the first sociology chair than any antipathy to the eugenic orientation itself. It is worth keeping in mind that even at the dawn of the 20th century neither biology nor sociology was a clearly defined field anywhere. 'Biology', a coinage of the first modern evolutionist, Jean-Baptiste Lamarck, challenged the classical way of thinking about 'nature' as consisting of animals, vegetables, and minerals as three equal modes of natural being. Instead Lamarck drew a sharp ontological distinction between living and non-living matter, effectively establishing the disciplinary boundary between biology and geology within the field that had been recognized from Aristotle to Linnaeus as 'natural history'. Fifty years later, Darwin was already taking this distinction for granted, as he tentatively posited the 'primordial soup' as part of an atheistic account of the transition from non-life to life. But Galton did not see his science of eugenics as an application of biology to human affairs. Rather, as Renwick makes clear, he saw it as an extension of political economy.

Nineteenth-century political economy was a quasi-normative discipline that treated everything as capital that could be inherited, accumulated, enhanced, and transmitted. In this context, eugenics made good on the bio-capital implications of the legal idea of 'inheritance', which can not only be taxed but also, so to speak, bred and farmed. Political economy had come into its own as the 'science of capitalism' once it junked the 18th-century French physiocratic idea that land – as proxy for nature – was the source of all value and focused instead on a conception of value as the conversion rate between forms of capital. At that point, political economy became committed to indefinite growth through ever more efficient substitutions of natural by artificial means of production, resulting in ever more productive form of capital. In this context, eugenics may be understood as extending the idea of increased agricultural productivity to what Darwin's French translator, Clémence Royer, called *puériculture*, which takes the idea of 'raising children' to a new degree of literalness. When the political economy backdrop to Galton's thinking is kept in view, then the route from late 19th-century eugenics to early 21st-century transhumanism is clear.

A key moment in this development was *The Principles of Political Economy* (1817) by David Ricardo, an English stockbroker who converted from Judaism to Unitarianism, the dissenting Christian sect dedicated to human self-empowerment that was associated with the radical chemist Joseph Priestley. Whereas Ricardo's older contemporary, Thomas Malthus (who himself was schooled in Priestley's curriculum at Warrington Academy), still believed that nature places an outer limit to productive growth, Ricardo abandoned that assumption, recognizing that even human labour would gradually lose its value through the introduction of more efficient mechanical substitutes. In this respect, attributing to Ricardo the 'labour theory of value' is a bit misleading, since for him the value of labour lies in the amount of it that is needed to make a commodity, *regardless of who or what delivers it*. Ricardo's 'labour' is not a constant but a variable – one normatively spun in the direction of 'least effort'. Before Ricardo, the labour theory of value (e.g. in Aquinas, Locke) had been tied to natural law theory, according to which human labour possesses absolute value, the source of the idea of 'just wage'. The quantity of labour was not abstracted from the labourer, as Ricardo had proposed to do. However, once Ricardo got his way, the door was opened to make all sorts of previously unseemly comparisons: e.g. one well-paid worker who dutifully works on schedule *versus* many poorly paid workers whose erratic performance collectively produces more.

While Ricardo himself appeared to believe (as many neo-liberals do today) that this situation provides an incentive for workers to acquire smarter skill-sets, if not commit themselves to 'lifelong learning', to keep up with the market, Karl Marx observed that Ricardian vision of capitalism seemed 'inexorable' only if the laws of political economy followed the path of least resistance to the capitalist employer. Ricardo's science of capitalism was in reality a science *for* capitalism. (Ricardo would try to regain the moral high ground by saying that Marx underestimates humanity's capacity for individual self-transformation.) To be sure, Marx was rhetorically effective in mobilising workers to organize themselves and speak with one voice, but it was at a cost. He effectively reverted to the labour theory of value associated with the natural law tradition, even though his own historical materialist metaphysical framework did not support it. Marx clearly did not want to turn back the clock to pre-capitalist days, since the efficiency savings encouraged by the capitalist mode of production was a necessary condition for a Communist paradise. Nevertheless, unlike Ricardo, Marx shared the natural law theorists' commitment to the integrity of the paradigmatically 'normal' human body, the legacy of which remains in the pejorative tinge attached to 'exploitation'. However, in practice, successful self-styled 'socialist' governments – be they in Scandinavia, Germany or Russia – operated in a more Ricardian spirit than Marx would have wished, one favourable to eugenics.

Galton's relevance to this debate is complex. While Galton questioned Ricardo's faith that individuals have the wherewithal to acquire new traits in response to changing market conditions, he refused to concede the finality of Darwin's Malthusian tendency to view these market shifts as expressions of natural selection that effectively decide who is fit to live. At the same time, Galton found Marx's counter-strategy to rely on an outmoded, even fetishised view of human labour (of the sort promoted by the medieval guilds) that failed to distinguish socially desirable traits from those who happen to bear them at a given time – a distinction Ricardo had clearly recognized. Galton's own strategy was to take the long view and try to persuade people that society's desirable traits are not normally well distributed across living individuals. Nevertheless, this suboptimal situation may be remedied by proactive policies designed to encourage and discourage births of certain sorts.

Precedent for this move could be found in Auguste Comte's mentor, Count Henri de Saint-Simon, the so-called utopian socialist who subsumed the human body under the category of 'property', the rational administration of which requires collective ownership and expert management. In that case, personal autonomy should be seen as a politically licensed franchise whereby individuals understand their bodies as akin to plots of land in what might be called the 'genetic commons', subject to all the rights and duties implied by the analogy. (An open question: Does this 'genetic commons' correspond to a racialized nation-state or a global human species?) The ultimate goal in this bio-capital utopia is maximum productivity – making the most out of one's inheritance. To be sure, 'irrational' (aka traditional) socio-economic barriers are likely to prevent some individuals – especially of poor backgrounds – from achieving this goal. And while wealth redistribution and egalitarian legislation might well address much of this problem in the short term, a more comprehensive long-term solution requires improving the capital stock of humanity itself. So goes the logic that leads to eugenics.

This last point is worth stressing for two reasons. One is contemporary: When faced with the shortfalls from the redistributivist and egalitarian policies that Western social democracies have pursued since the 1960s if not earlier, it is nowadays common for left-leaning, biologically minded thinkers to declare – as Darwin himself might – that there are definite limits to how much people can be changed. Indeed, in his 1999 manifesto, *A Darwinian Left*, Peter Singer went so far as to advise his fellow leftists to ditch Marx for Darwin. Whatever else one might wish to say about eugenics, it did not give up so easily – or more precisely, it had a more consistent faith in the import of new knowledge (aka 'basic research') for future policy-making. The other reason to elucidate the logic behind eugenics is to dispel a pervasive historical stereotype. Because eugenics continues to be closely associated with the totalitarian regimes of Nazi Germany and the Soviet Union, Galton's science is often seen as aiming for policy outcomes much more quickly than could (or would) be achieved by normal democratic processes. However, in the British soil where eugenics first took root, its most outspoken advocates – Sidney and Beatrice Webb – identified themselves as 'Fabian socialists', in the spirit of Fabius, the Roman general who refused to act impulsively against Hannibal in the Punic Wars but nevertheless won in the end. In other words, eugenics was supposed to provide a blueprint for basic research in the social sciences with a rather long-time horizon, comparable to the experimental turn that had enabled the natural sciences to break with the natural history tradition over the previous two centuries.

Recall that the people normally taken to be the founding fathers of the social sciences (excluding psychology) believed that we either already knew enough about the human condition to now focus on its political implications or, if our basic knowledge was still lacking, we would proceed more systematically but in largely the comparative-historical mode of traditional humanistic scholarship. The former category included Comte, Mill, and Spencer, whilst the latter included Durkheim and Weber, with Marx believing in a bit of both. Galton's eugenics was arguably the first discipline to offer a clear statement of a basic research programme for the social sciences that had something like the character and dimensions in terms of which funding agencies think about such matters today – that is, a strong theoretical framework operationalized in terms of clear methodological strictures that enabled the collection and analysis of a wide range of original data. Put this way, it should come as no surprise that Otto Neurath, the sociological founder of logical positivism, was Galton's German translator. Indeed, eugenics would not have been such an easy target for censure, had it not set its own scientific standards so high – something for which the field has yet to be given due credit.

Moreover, as Renwick (2011) shows in his current work the epistemic significance of eugenics was not lost on the father of the British welfare state, the economist William Beveridge. In 1930, as director of the LSE, Beveridge hired the experimental biologist Lancelot Hogben to establish a department of 'social biology' that would provide a 'natural basis for social science'. But once again, it would be misleading to see in this project the sort of biological imperialism that, say, characterized E.O. Wilson's 'sociobiology' of the 1970s. On the contrary, Renwick shows that Beveridge and Hogben saw the uncritical extension of animal-based studies to human populations as profoundly unscientific, making for capricious policy. As Hogben wittily put it, social biology needs to be less about 'the sterilisation of the unfit' than 'the sterilisation of the instruments of research before operating on the body politic'. In his brief and unhappy tenure at the LSE, Hogben managed to launch a sophisticated survey of 4000 twins of school age in the London area to examine in some detail the relationship between heredity, environment, and intelligence – with an eye to checking the validity of psychological testing. And while a failure in his own terms, Hogben nevertheless did train David Glass, who went on to become the doyen of British quantitative sociologists in the postwar era. As a passing shot of this era, it is interesting to note that the main difficulty for Beveridge in persuading his fellow LSE economists – not least Friedrich Hayek – of the need for social biology was simply convincing them that the social sciences needed 'basic research' at all.

The appointment of L. T. Hobhouse to the LSE's first sociology chair, effectively making him the founder of sociology in the UK, may have been the strategically best appointment in 1907. Without denying this basic judgement, Renwick puts the original contenders on equal footing, as they might have appeared back then, hinting at the alternative histories of social science that would have resulted. In particular, he normalises the prospect of a eugenics-based social science, which remained very much alive into the 1930s, despite the bad politics that had already come to be associated with it. Today, with rapidly advancing frontiers in biotechnology that are wreaking havoc on the traditional disciplinary structures of both the social and the biological sciences, eugenics agendas are being advanced in everything but name. This book provides a clear and sober route to repatriating these discussions in the history of the social sciences, where they belong.

Steve Fuller,
Auguste Chair in Social Epistemology,
University of Warwick, UK.

Introduction

In 1985 the prominent American sociologist Edward Shils, Distinguished Service Professor of Sociology at the University of Chicago and winner of the prestigious Balzan Prize, looked back at the ways his discipline had developed since its inception a little over 150 years earlier. 'The sober attempts of a small group of dourly upright reformers and administrators in the nineteenth century to describe in a reliable way the real "condition of England" were among the first of their kind', he wrote. The reason was that 'for the first time men and women sought to arrive at a judgment of their own society through the disciplined and direct study of their fellow citizens'. However, Shils went on, after a 'great surge', which lasted from the Poor Law Commissioners of the 1830s until the surveying work of Charles Booth and Beatrice Webb in London during the 1880s and 1890s, 'British sociological powers seemed to exhaust themselves'. In early twentieth-century France and Germany, Shils declared,

> powerful and learned minds thought about the nature of society and tried to envisage modern society within the species of all the societies known to history. In America, sociologists busied themselves in villages and in the city streets, carrying on the work of Booth, finding illustrations of the ideas of [Georg] Simmel, [Ferdinand] Tönnies, and [Emile] Durkheim and developing under the guidance of Robert Park, a few of their own. In Britain, however, for nearly fifty years, while social anthropology and economics flourished as in no other country, sociology gathered the soft dust of libraries and bathed in the dim light of ancestral idolatry.[1]

Far from being specific to Shils, this view of British sociology is emblematic of a general historiographic trend, which suggests that the UK failed

1

to develop a sociological imagination during the late nineteenth and early twentieth centuries. Indeed, for many scholars who have written about the history of sociology, there is little to say about the theoretical, rather than empirical, dimension of the subject in Britain before the emergence of Anthony Giddens during the late 1960s and early 1970s. As this book demonstrates, though, this perception of the discipline's past is not only misleading, it also conceals a history that illuminates a whole range of current debates about how to understand and reform society.

At the heart of this book is a new account of the intellectual origins of British sociology, which is focused on a series of late nineteenth- and early twentieth-century discussions that have been overlooked by scholars such as Shils. During the course of those debates, which culminated between 1903 and 1908 with the creation of the UK's first organization, university chair, and journal for sociology, a number of different visions of the burgeoning discipline's future were put forward. However, only one of those visions, that of the former Oxford philosophy don L. T. Hobhouse, who was appointed Martin White Professor of Sociology at the London School of Economics in 1907 and editor of the *Sociological Review* in 1908, emerged from those discussions with the power to direct sociology in its institutional setting. By charting the emergence of sociology in the UK from the mid-1870s through to Hobhouse's double appointment during the first decade of the twentieth century, this book recovers the frequently ignored context and often forgotten content of the founding British debates about sociology and explains why they concluded with the selection of Hobhouse, rather than any of his rivals, for the discipline's first and most important jobs.

In so doing, this book traces how the British understanding of the identity of sociology – what it should strive for, what methods are appropriate to it, and how it fits with the rest of science and culture – was developed in response to a specific question: how should sociology, as the general science of society, be related to biology, as the general science of life? What is shown is that, by choosing Hobhouse, those who laid the intellectual and disciplinary foundations for British sociology also chose to keep biology and sociology separate – a decision that had enormous consequences for the field's identity. Yet despite the fact that this issue was not only of such importance to early British sociology but has also re-emerged in recent years as a subject of debate in sociology, it is seldom mentioned in the history that sociologists tell about their discipline.

The origins and development of an independent science called 'sociology'

From the writings of the ancient Greeks, through the eighteenth-century Enlightenment's aspirations for a 'science of man', to the social and behavioural sciences of the twentieth and twenty-first centuries, Western thought has always featured writings that have aimed to understand man and society in a systematic way. However, there was no such thing as sociology before the nineteenth century. As is well known, the word 'sociology' was coined by the French positivist philosopher August Comte in the 1830s as the name for the science he believed would enable humans to comprehend society in the same way that mathematics, physics, chemistry, and biology had made it possible for them to understand the natural world. Yet after his writings about the prospect of replacing existing theological beliefs and arrangements with a 'religion of humanity', which the English naturalist and public defender of Darwin, T. H. Huxley, once famously described as 'Catholicism *minus* Christianity', sociology possessed such radical social and political connotations that few outside of Comte's loyal group of followers were prepared to openly identify themselves with the idea.[2] It was therefore not until other writers, in particular the English philosopher of evolution Herbert Spencer, had separated sociology from its original Comtean context that it became the subject of serious, mainstream scientific debate in Europe and North America.

Of the thinkers whom sociologists now believe contributed most to the intellectual foundations of their field, it is a group known as the 'classical' generation, with which thinkers such as Emile Durkheim, Ferdinand Tönnies, Max Weber, Wilfredo Pareto, Georg Simmel, and George Herbert Mead are associated, whom sociologists consider the most important. The reason is that the classical generation, unlike Comte and Spencer, provided a definition of sociology and its subject matter that has informed the work of sociologists ever since. For the classical generation, sociology was not simply the scientific study of man but the more precise practice of studying a realm of phenomena, known as the 'social', which they claimed has an existence beyond the individuals and political organizations that had been the subjects for their predecessors. Moreover, in pointing to this class of phenomena, the classical generation also argued that its study required a new and specific set of analytic tools and methods. In this sense, as Durkheim put it in 1895 in *The Rules of Sociological Method*, sociology was not 'the appendage of any other science' but was instead 'a distinct and autonomous science'.[3]

As Shils suggested, the consequence of these arguments about the independence of sociology from other fields was the formalization by a number of thinkers of a new way of looking at society. The overriding concern of what C. Wright Mills later called the 'sociological imagination' was to show how human action is constrained by the traditions, ideals, and formal arrangements of the societies in which it takes place.[4] For example, in *Suicide*, which was first published in 1897, Durkheim sought to establish that suicide is not just a psychologically motivated act but also a social phenomenon that is related to the degree of social solidarity present in the society in which it occurs. Moreover, using his 'ideal type' methodology, Weber laboured to provide an account of how human action, despite its subjective meaning to individuals, can be classified into a number of distinct types, such as value-rational and instrumental kinds.[5] Although the classical sociologists applied such approaches to societies throughout history, they were primarily interested, as Shils observed, with explaining what made the industrial societies in which they lived different from the social forms that had preceded them. Whilst writers expressed this concern in many different ways, it was perhaps most famously summed up by Tönnies in 1887 in *Gemeinschaft und Gesellschaft*, known as either *Community and Association* or *Community and Civil Society* in English, which contrasted the traditional relationships in static and small-scale communities with the informal contracts of the rapidly changing societies in the modern world.[6]

The achievements of the classical generation were not limited to intellectual insights alone, though. By establishing journals, professional organizations, and university departments, the classical sociologists founded a basic disciplinary framework for sociology that, in many cases, still exists today. In France, for example, Durkheim not only helped establish sociology as a university subject through his teaching at the University of Bordeaux and then at the Sorbonne in Paris, but also participated in the founding of *L'Année sociologique* – Europe's first journal for sociology. Furthermore, in the USA, which took the disciplinary process further than any other country in the late nineteenth and early twentieth centuries, figures such as Franklin Giddings – America's first professor of sociology – and Albion Small – the first editor of the *American Journal of Sociology*, which commenced publication in 1895 – established the institutions that would serve as the basis for sociology's expansion after World War I. Indeed, the case of the department of sociology at the University of Chicago, which Small helped found during the 1890s, is a model of institution building that has interested scholars ever since.[7]

In writing the history of their discipline, however, sociologists have woven these general points and specific events into a single and greatly expanded narrative that has not only glossed over uncomfortable details but also deepened the subject's intellectual roots – a fact that is to be expected from scholars working in a field that is less than 150 years old. For example, few histories of sociology engage with the fact that Weber, who never held a professional sociology post, actively avoided the label of sociologist until around 1910 when he had not only already established his methodological principles and theoretical framework but also published works, such as *The Protestant Work Ethic and the Spirit of Capitalism,* for which he has subsequently become most renowned.[8] Furthermore, and highlighting the importance sociologists have attached to the identity of their field as the science of industrial life, Karl Marx has come to be embraced as a sociological thinker, even though he neither called himself a sociologist nor was identified as one during his own lifetime. Indeed, such is his importance to sociology's sense of its own identity that it has become commonplace to see classical sociology as a response, from Weber in particular, to Marx's historical materialism.[9] Yet whilst the writing of sociology's history has involved the embracing of thinkers whose status as sociologists was somewhat ambiguous during their own lifetimes, it has also seen the exclusion of others. In particular, the received history of sociology has not only overlooked the participation of the British in the late nineteenth- and early twentieth-century debates about the subject but also dismissed those contributions as largely inconsequential.

The trouble with British sociology

In common with the USA and its European neighbours, early twentieth-century Britain was a place where sociology captured the imagination of those interested in being scientific about society. Indeed, between 1903, the year of Herbert Spencer's death, and 1908, a society, a university chair, and a journal were all established for sociology in the UK. Yet, as historians and sociologists have often observed, that flurry of activity did not translate into any substantial institutional gains until after World War II. Moreover, not a single person involved with the process of founding sociology as discipline in Britain has ever been widely considered worthy of a place alongside the greats of the field's classical canon. As a consequence, the historiography of British sociology has been divided between two groups. On the one hand, there are those, like Shils, who have been genuinely dismissive towards what happened in Britain during sociology's classical age. On the other, there are scholars who have sought to defend the UK from the claim that its failure to develop sociology along the same lines as the rest of Europe and America represents a deficiency of some kind. For almost everyone who has written on the subject, though, there has been an acceptance that there is something distinctive about late nineteenth- and early twentieth-century British sociology that requires an explanation.

Amongst the most important critiques of British sociology, and one that helped shape a whole generation's perception of the field, was that of the American sociologist Talcott Parsons, whose theory of structural functionalism dominated sociology during the middle decades of the twentieth century.[10] A key figure in the introduction of Weber's work to English-language audiences during the 1930s, Parsons extolled the virtues of the European classical theorists and presented sociology as a project that had been built on their insights.[11] Perhaps surprisingly, Parsons had actually been introduced to the work of a number of the classical thinkers when he was studying at the London School of Economics during the late 1920s. However, it was not L. T. Hobhouse, then still the UK's only holder of a chair of sociology, who had been responsible for drawing Parsons' attention towards the classical canon. Instead it was through the historian R. H. Tawney that Parsons learned of Weber and through the anthropologist Bronislaw Malinowski that he first came into contact with Durkheim's writings.[12] Comparatively unimpressed by the style of sociology that Hobhouse was offering at the LSE at the time, Parsons concluded that British sociologists were unwilling to grasp the abstract theories of their European contemporaries, who he argued in *The Structure of Social Action*, his widely influential book of 1937, were central to the entire enterprise of sociology. In fact, for Parsons, the British were too deeply wedded to an empirical and individualist style of social thought, which he identified as a tradition running from Thomas Hobbes and John Locke in the seventeenth century through to the economics of his own time, to ever embrace sociology proper.[13]

Of those who followed in Parsons' footsteps, the English Marxist historian Perry Anderson, who despaired at the UK's failure to produce its own Durkheim or Weber, stands out as the scholar who has offered the most direct attack on the British sociological tradition. As a subscriber to the belief that classical sociology emerged from a dialogue with Marx, Anderson argued in a much discussed article entitled 'Components of the National Culture', which was published in 1968, that the key to understanding why British sociology did not develop along the same lines as it did in Europe is the UK's long-standing political and social stability.[14] According to Anderson, late nineteenth-century France and Germany were susceptible to classical sociology and the grand theorizing it represents because radical leftwing political thought found a place in those countries after the massive upheavals that were caused by invasion, revolution, and unification. In Britain, though, the foundations of the state and government were stable and, as a consequence, there was no platform on which to build sociology. However, Anderson went on, this stability and apparent absence of sociology was not a good thing. On the contrary, Britain's failure to produce a classical sociology was, he argued, a damning indictment of its intellectual culture.

Since the 1960s, numerous scholars, many of them sociologists, have attempted to defend British sociology and, by extension British intellectual life, against the charges levelled at it by Parsons, Anderson, and others. One popular response, which was reflected in Shils' take on the subject, has been to emphasize that despite not having given the world a substantive body of theory, the UK has nevertheless contributed to sociology in other important ways. Specifically, as scholars such as Lawrence Goldman, Raymond Kent and Philip Abrams have pointed out, Britain has always possessed a strong tradition of empirical and statistically orientated social research, which can be traced from the founding of the Statistical Societies of Manchester and London in the mid-1830s, through the late nineteenth- and early twentieth-century work on urban poverty by investigators such as Charles Booth in London and Seebohm Rowntree in York, right up to the late twentieth century, when British sociology came to be characterized by 'studies' rather than theory.[15] For this reason, many scholars have emphasized Britain's strong commitment to the empirical foundations of social science, which, if we eschew Parsons and Anderson's normative claims about the direction that British sociology should have taken, provides a positive history of sociology in the UK.[16]

However, shifting attention towards the undoubtedly strong empirical tradition in British social science has never provided an effective riposte to the claim that the UK has been an intellectual wasteland in comparison to its European neighbours. As a consequence, scholars have frequently explored ways to demonstrate that whatever went wrong with British sociology it was by no means a deficiency of the intellect. Indeed, according to a number of historians and sociologists, the conclusion to be drawn from the failure of sociology to become a significant institutional proposition during the first half of the twentieth century is not that it was absent from British culture but rather that it was dispersed throughout it. For Philip Abrams and Geoffrey Hawthorn, for example, sociology was at a disadvantage in late nineteenth- and early twentieth-century Britain because of politics. Not only did the British state embrace social, or vital, statistics, Abrams and Hawthorn have argued, but the political culture was gradually shaped by a liberal-social political consensus, which was premised on a sophisticated understanding of the structural causes of poverty.[17] Given that the natural audience for sociology was social reformers, who saw social science as a tool for social and political change, British sociology suffered, Abrams and Hawthorn have claimed, because too many people, including Edwin Chadwick and Beatrice and Sidney Webb, were able to realize their ambitions through the machinery of government rather than academia.

For a number of scholars, the problem of viable alternatives to sociology in late nineteenth- and early twentieth-century Britain was not limited to politics, though. Indeed, for Abrams, there were a number of other disciplines, such as social anthropology, which flourished in the UK during the late nineteenth and early twentieth centuries, which can be seen as having absorbed thinkers who might otherwise have pursued sociology.[18] Yet of all the disciplines Abrams suggested as having a role to play in the history of British sociology, it is philosophy that has received the most attention from other scholars – a fact that should not be surprising given that Hobhouse started his career as a philosopher. As Stefan Collini and, most recently and extensively, Sandra den Otter have shown, sociology and philosophy, which was then dominated by an idealism related to the work of Immanuel Kant and G. F. W. Hegel, intersected at a number of different points in late nineteenth- and early twentieth-century Britain. Whilst a number of sociologists, including Hobhouse, drew on aspects of idealist philosophy in their work, a prominent group of philosophers, such as Bernard Bosanquet and D. G. Ritchie, attacked sociology because they did not believe society could be treated within the framework of scientific naturalism. For Bosanquet, Ritchie, and others, human action could not be understood in materialist terms and, as a consequence, they objected to the idea that it was possible to be scientific about society. In this sense, late nineteenth- and early twentieth-century British sociology was the site of an important and scholarly discussion about the most appropriate away to understand society and therefore whether it is the sciences or the humanities that are best placed to interpret it.[19]

The result of these efforts to recover the context in which people talked about and related to sociology in late nineteenth- and early twentieth-century Britain is that it is now by no means the case that one can casually claim, as scholars such as Shils, Parsons, and Anderson once did, that the UK was, and to some extent still is, an intellectual poor relation of Europe and the USA. Building on the important studies by Goldman, Collini, den Otter, and others, this book contributes towards the ongoing spirit of revisionism by returning to a formative set of British debates about sociology and arguing for two general points relating to the historiography of the field as a whole. The first is that we should follow the example of revisionist scholars and reject the argument that British sociology's institutional fortunes during the twentieth century can be put down to an absence of ideas about the structure and nature of society. However, the second point is that we do not need to compensate for any perceived deficiencies in the standard of sociological debate in late nineteenth- and early twentieth-century Britain by appealing to what was going on in other disciplines. There was a rich, complex, and theoretically informed discussion in late nineteenth- and early twentieth-century British sociology; but, as this book shows, that debate was largely concerned with a set of issues that seldom appear in accounts of how the current identity of sociology came to be.

Biology and British sociology

The presence of not only biological ideas but also biologists in the late nineteenth- and early twentieth-century British debate about sociology is something that a number of scholars have recognized.[20] Indeed, as early as 1948, in what still stands as the most comprehensive survey of sociology from its origins to the mid-twentieth century, the American historian Harry Elmer Barnes commented on the UK's strength when it came to making connections between the sciences of society and the sciences of life.[21] It is therefore surprising that very few scholars have ever paid any attention to the part biology played in laying the foundations of sociology in Britain. In fact, of those who have considered the subject at all, only R. J. Halliday, in a 1968 study of the major groups that formed the Sociological Society, has ever taken the issue seriously enough to suggest that biology provided at least some of the reference points when it came to the disputes that divided British sociology during the early twentieth century. However, since Halliday suggested closer attention should be paid to how members of the British sociological movement related biology and sociology, little work has actively engaged with the issue.[22] This book demystifies the place of biology in late nineteenth- and early twentieth-century British sociology and, in so doing, it argues that if we want to understand why British sociology became a Hobhousean enterprise then we have to appreciate the significant part that biological ideas played in making it so.

Over the course of six chapters, which are divided into three parts, this book explores the British debate about the methods, scope, and aims of sociology from the perspectives of the three men who arrived at the Sociological Society in the early twentieth century with the most comprehensive and significant proposals for how sociology should be practiced: the eugenicist and biostatistician Francis Galton; the biologist and sociologist Patrick Geddes; and the former Oxford philosophy don Hobhouse. Exploring how each of their programmes was based on a different understanding of how to relate sociology and biology, this book restores the debates that took place at the Sociological Society between 1903 and 1907 to their correct intellectual context and recaptures what was at stake for British sociologists when they decided to follow the path laid by Hobhouse rather than Galton or Geddes. Indeed, this book highlights the significance of what happened at the Sociological Society by demonstrating how very different sociology would have been in the UK had it been built on the biosocial foundations proposed by both Galton and Geddes.

Chapter 1 begins by examining the events that first pushed sociology into mainstream British scientific discussions. Whilst received histories of sociology suggest a straight line can be drawn from Comte's coining of the word 'sociology' in the 1830s to the discipline that took shape in the twentieth century, Chapter 1 argues that the origins of British sociology are to be found elsewhere: in the rise and fall of classical political economy. For much of the nineteenth century, the classical form of political economy had dominated British social science. However, as Chapter 1 shows, confidence in the doctrines of classical political economy had declined to such an extent by the late 1870s that there was a hard-fought campaign to close Section F, the social science branch of the British Association for the Advancement of Science (BAAS). Although one well-known consequence of the late nineteenth-century revolt against classical political economy was the emergence of modern economics, Chapter 1 demonstrates that another was the rise of sociology. Indeed, what is shown is how the Irish political economist J. K. Ingram, responding to both the decline of classical political economy and the related attack on social science at the BAAS, first made sociology a talking point in British social science through his presidential address to Section F in 1878.

Chapters 2, 3, and 4, which comprise the second part of the book, then examine how Galton, Geddes, and Hobhouse developed their programmes for sociology during the late nineteenth century. The common theme of these chapters is the effort of each thinker to be scientific about society but in an intellectual environment that had been shaped by two things: the issues raised during the debate about classical political economy and the impact of evolution on social thought after the publication of Charles Darwin's *On the Origin of Species* in 1859. Whilst it has long been assumed that the most biologically inclined of the programmes that emerged from that process was Galton's eugenics, we will see in Chapter 2 that there were actually few substantive links between his work on human heredity and the practices of late nineteenth-century biologists. Motivated by a political programme that was shaped by his objections to the assumptions that classical political economists made about human nature, Galton developed eugenics as a rigorously statistical science through which he strived to convince people that the answers to social questions could be found through a closer study of the laws governing inheritance. However, as a number of high-profile biologists and statisticians sought to incorporate Galton's insights into their work from the 1890s onwards, eugenics was transformed into a social science with strong connections to the latest biological thinking. Consequently, Chapter 2 argues, when Galton appeared before the Sociological Society in the early twentieth century, he was proposing that sociologists pool their intellectual and practical resources with their colleagues in biology.

As Chapter 3 shows, the programme for British sociology that actually owed the most to biology was the one developed by Patrick Geddes. Once a student of the anatomist T. H. Huxley, Geddes had been set for a career as a biologist in the late 1870s when he decided to immerse himself instead in the discussions that had been ignited by Ingram's 1878 address to the BAAS. By examining Geddes' work in biology, his responses to the debate about classical political economy, and his attempts to establish sociology within his home city, Edinburgh, Chapter 3 traces how his programme for sociology, which he called 'civics', was the product of two concerns: Geddes' desire for sociology to be an independent science and his conviction that it needed to engage with biology. Moreover, in charting Geddes' efforts to balance these demands, Chapter 3 highlights the previously undocumented ways that his sense of how to do so was shaped by the evolutionary philosophy of Herbert Spencer – a source of inspiration that historians of sociology unfamiliar with the complexities of the late Victorian debate about evolution might find surprising. In this sense, it will be shown how Geddes, like Galton, arrived at the Sociological Society with a programme that required sociologists to incorporate particular aspects of biology into their practices.

In Chapter 4, we will then see how Hobhouse spent the late nineteenth and early twentieth centuries researching and writing a multi-volume project that began in late 1880s Oxford and concluded in 1906 with the publication of *Morals in Evolution* – a statement of how sociology should be done and what it should strive for. Beginning his career as a philosophy don, Hobhouse had been drawn into the debate about sociology by his belief, contrary to a number of his illustrious contemporaries in philosophy, that theories of evolution had important implications for the way that philosophers thought about their subject matter. As with Geddes' attempts to relate society and evolution, the spectre of Spencer's Synthetic Philosophy loomed large over Hobhouse's project. However, whilst modelling the structure of his work on Spencer's system, Hobhouse pursued his research with the determination to overcome the weaknesses that commentators had come to identify with Spencer's account of evolution. In so doing, Hobhouse formulated an agenda for sociology that made human agency a key part of the evolutionary process and thus social change. Yet, as Chapter 4 makes clear, whilst Galton, Geddes, and Hobhouse all shared this normative understanding of sociology, only Hobhouse was convinced that its realization was dependent on freeing sociology from biology.

Building on these analyses, Chapters 5 and 6 then provide a new account of what happened at the Sociological Society in the lead up to Hobhouse's appointments in 1907 and 1908. Whilst using the new context provided by Chapters 1 through 4 to throw fresh light on what was said and who took part in those debates, Chapters 5 and 6 also utilize correspondence between the major figures in the British sociology movement, almost all of which has been unknown to historians of sociology, to explain exactly what happened at the Sociological Society between its founding in 1903 and Hobhouse's appointments. In so doing, Chapters 5 and 6 not only dispel myths about how Hobhouse came to be selected for the chair at the LSE but also reveal for the first time that there was a plan hatched by Victor Branford, Geddes' closest associate, to make Hobhouse the leader of British sociology. As the correspondence and reports of the Sociological Society make clear, the reason it was Hobhouse, rather than Galton or Geddes, who was given the opportunity to shape sociology in the UK was the Society's difficulties with Galton and Geddes' calls for sociology to be linked to biology.

These themes and arguments will then be brought together and considered more closely in the conclusion. However, rather than evaluating what went on at the Sociological Society in terms of its supposed status as a precursor of the subsequent failure of sociology in Britain, the conclusion will ask a different question: what might British sociology have looked like had it been put in either Galton's or Geddes' hands? Whilst the conclusion takes this counterfactual turn to avoid the problem of reading later institutional difficulties back into the debates of the early twentieth century, it also does so to underscore an important point: that the events considered in this book mattered when they took place and still matter now.[23] By pointing to the ways that sociology would have been different on Galton's or Geddes' watch, the conclusion explains not only what those who took part in the debates at the Sociological Society understood to be at stake when Hobhouse was appointed but also how the effects of that appointment can still be felt in the discussions that are taking place in British sociology today. Indeed, it is argued that it is only by reconnecting with the reasons behind Hobhouse's appointment that we can make sense of what is at stake right now when sociologists such as W. G. Runciman call for their colleagues to 'forget their founders' in favour of a Neo-Darwinian paradigm.

History of science and the history of sociology: wider aims

As this high-altitude sketch suggests, this book also has a number of aims that lay beyond the recovery of the context in which debates about sociology took place in late nineteenth- and early twentieth-century Britain. Indeed, whilst the question of why Hobhouse was appointed the UK's first professor of sociology and editor of its first sociology journal might ostensibly seem to be of interest to a small group of scholars, this book endeavours to show how the answer is actually relevant to a range of debates from both sociology and history of science. Although these wider issues and discussions intersect at a number of different points, they can be divided, broadly speaking, into three main groups.

The first set of issues concerns the relationship between the history of sociology and history of science more generally. Simply put: why should historians of science, who have paid little attention to the history of British sociology before, be interested in the contents of this book? As the preceding sketch made clear, one reason is that the history told here involves well-known thinkers engaging with questions that had yet to be monopolized by any one field. The British debate about sociology is therefore important because it provides a crucial but under-appreciated context for issues and developments that historians of science have previously studied through the lens of other sciences and their histories. For example, as Chapters 2, 5, and 6 show, the debate about sociology provides a significant point of reference when it comes to understanding how and why Francis Galton developed his ideas about human heredity in the ways he did. Indeed, as Chapter 5 shows, it is only by taking that debate and its impact on Galton's research into account that we can explain why he chose to launch his campaign for eugenics to be taken up as a national political concern at an organization called the Sociological Society – a fact that few scholars have commented on before. In this respect, and building on a point that Steven Shapin has made in a number of recent reviews, this book speaks to an important problem in history of science scholarship: the fact that historians of science, despite their enthusiasm for contemporary sociology and social theory, frequently neglect the social sciences when it comes to the contexts in which their subjects of study were working.[24]

The second set of general issues that this book tackles concern the relationship between history of science and the sciences themselves. As Peter Dear has pointed out, although history of science and the sciences are now two largely separate enterprises, they were closely entwined as recently as a century ago.[25] Whilst the reasons for this shift have yet to be fully documented, one of its consequences has been a weakening of the link that once existed between scientific ideas and the understanding of their historical context. For a great many historians, as well as philosophers and sociologists, of science, this growing separation is of great concern and the question of how to bring science and the study of science closer together has been the subject of a growing discussion.[26] This book joins that discussion by using the case of British sociology, where questions about the place of biology in social explanation have recently returned to the agenda, as an example of how historical investigation can constitute a meaningful contribution to current scientific debates. Indeed, as the conclusion makes clear, the recovery of the significant differences between Galton, Geddes, and Hobhouse's visions of sociological practice is a not only significant but also necessary when sociologists are calling for greater attention to be paid to Geddes' programme for sociology.

This conception of how historical scholarship can be related to contemporary scientific discussion is connected to the third and final set of general issues that this book addresses: how historians conceptualize science and write its history. As James Secord helpfully summarized in *Isis* in 2004, the writing of the history of science during the past two decades has been characterized by a shift towards a conception of science as a practice.[27] Motivated by a desire to redefine science as something that is moulded by more than just theories, scholars have made a whole range of subjects, from book publishers to instrument makers, a part of our understanding of what science is. However, one important, and perhaps unintended, consequence of these developments has been an overall privileging of material culture in history of science at the expense of what have come to be derided as 'elite' ideas. This state of affairs can be gleaned from the contents of history of science journals, which have come to be dominated by an ever-growing body of constructivist studies that focus on issues such as laboratory skills and the communication of evidence to audiences – a collection of interests that are often grouped under the heading 'making knowledge'. Yet as Jan Golinski and Jonathan Topham have both argued, thinking is clearly an important part of scientific practice.[28] For this reason, this book focuses on the connections between thinking and doing in scientific practice and, in so doing, aims to transcend the dichotomy between intellectual and practical activity that has bedevilled the contemporary historiography of science.

In terms of the six chapters that follow, the most obvious and well-known example of a close relationship between ideas and action comes in Chapter 2 with Galton's efforts to develop a set of statistical tools that could be used to link social progress and human heredity. However, in Chapters 3 and 4, where Geddes and Hobhouse's programmes for sociology are shown to be indebted to their appropriation of Herbert Spencer's evolutionary philosophy, we will also see how writings that many scholars have dismissed as irrelevant to the concerns of practicing scientists actually shaped a range of activities in both the physical and intellectual worlds. From Galton's endeavours to formulate methods that would accurately capture information about human intelligence and form, through the decisions that Geddes made during his research on what we would now call symbiosis, to the way that Hobhouse designed a set of experiments to test the reasoning abilities of animals, those involved in late nineteenth- and early twentieth-century British sociology were all engaged in an effort to develop and execute programmes that were driven by specific sets of ideas. In this sense, what this book aims to show is why we should not lose sight of the fact that science is a cognitive activity and that studying that activity is a crucial part of understanding the direction of science in the past, present, and future.

However, in order to make this argument, as well as the others that have just been outlined, we must first return to the late nineteenth-century debate about classical political economy that helped bring Galton, Geddes, and Hobhouse together at the Sociological Society during the first decade of the twentieth century.

Part I

1
Political Economy, the BAAS, and Sociology

At the British Association for the Advancement of Science's (BAAS) 1878 meeting in Dublin, J. K. Ingram (1823–1907), a founding member of the Irish Statistical Society and the president of the Statistical and Social Inquiry Society of Ireland, addressed the Economic Science and Statistics Section (Section F) as president. 'There is…no duty more incumbent in our day', he announced, 'than that of recognising the claims of Sociology, whilst at the same time enforcing on its cultivators the necessity of conforming to the genuine scientific type.'[1] The reason sociology was so important, Ingram explained, was because of a debate that had engulfed British social science. As his audience knew, the criticism of one social science in particular, political economy, had been so fierce that Section F had recently been threatened with closure. Ingram was therefore calling for minds to focus on sociology because he believed that this new field could end those controversies and, in the process, inaugurate a new era for social science. By tracing the origins and escalation of the debates that inspired Ingram to make this claim, this chapter explains how sociology was first established as a talking point in the UK. In so doing, the chapter shows why it was in the late 1870s and through a debate about political economy that sociology emerged in Britain as a science that it was hoped could guide not just social investigation but social action as well.

Welcomed by *The Times* as a 'thorough and unsparing…discussion by so competent a hand', Ingram's presidential address to Section F has been recognized frequently since as a landmark in the history of British sociology.[2] Most notably, Philip Abrams included an edited version of the speech amongst the collection of documents that make up the second half of his history of the discipline.[3] The reason such importance has been attributed to Ingram's address is that it represents the moment when sociology went from being the intellectually dubious pursuit of Comtean positivists, who were tainted by the 'religion of humanity', to something that was part of mainstream scientific discussion. To be sure, Ingram, himself a Comtean positivist, though not yet fully out, was not solely responsible for this change of attitude as a number of writers, including the philosopher of evolution Herbert Spencer, had begun using the term 'sociology' in the early 1870s.[4] However, on account of when and where it was delivered, Ingram's speech not only drew attention to sociology in a way that those earlier efforts did not but also marked the beginning of a debate about the scope, methods and aims of sociology, which would reach its climax between 1903 and 1907 with the establishment of the Sociological Society.

Yet despite the recognition of Ingram's speech as a turning point for British perceptions of sociology, historians have seldom studied the context in which it was delivered. More specifically, whilst scholars have recognized that Ingram made a general plea for the need of a science called sociology, they have failed to explain why it was a debate about political economy that enabled him to successfully make the case for sociology when and where he did: at Section F of the BAAS in 1878. By exploring these issues, this chapter recovers both the intellectual origins of the British debate about sociology and establishes the concerns that would shape discussion of the subject in the late nineteenth and early twentieth centuries. Indeed, by showing why it was at the BAAS that sociology was established as a major talking point in the UK, this chapter relates the emergence of sociology to the redrawing of the intellectual map of British social science – a process that was catalysed by the decline of the reputation of 'classical' political economy, which Ingram believed sociology should replace.

Beginning with an outline of the key elements of what has come to be known as its 'classical' system of thought, this chapter explains why, by the mid-1870s, political economy occupied a position of such popularity in Britain that Beatrice Webb was brought to describe it as 'the only social science we English understood'.[5] However, whilst outlining how classical political economy evolved to become the UK's leading normative social science, and one that was embraced by intellectuals, businessmen, and politicians alike, the chapter also pays attention to its critics. Indeed, as the second section shows, a site of particularly fierce debate over the virtues of classical political economy was the BAAS' Section F, which was founded in the 1830s with the aim of challenging the field's dominant methods and rationale. However, as the third section shows, those efforts were long unsuccessful and it was not until the 1860s that critics were able to dent classical political economy's lofty reputation, which, as the fourth and fifth sections show, had a double-edged effect. Whilst the classical system of thought went into what would turn out to be a sudden and rapid decline, one consequence of that shift was a re-igniting of the debate about political economy in Section F. Resituated thus, in the context of Section F's long-standing and newly resurgent debate and criticism of classical political economy, Ingram's 1878 call for social scientists to concentrate on the potential of sociology to succeed where political economy had failed, reconnects with its polemical intent.

Although there are a number of intersecting narratives throughout this chapter, there is one figure who binds them all together: William Gladstone's first Chancellor of the Exchequer, Robert Lowe (1811–92). Described by his first biographer as 'a profound student of the writings of Adam Smith and [David] Ricardo', Lowe's devotion to bringing the rationality of the *laissez-faire* principles associated with those thinkers to bear on the organization of the state made him widely representative of the mid-Victorian integration of political economy, business and government.[6] As well as placing theories from political economy at the fore of parliamentary debate, Lowe unintentionally began a chain of events in 1876, the centenary year of Smith's *Wealth of Nations*, that culminated with Ingram's address. Indeed, Lowe was so intimately associated with political economy that he was compelled to engage publicly with Ingram over his calls for sociology to replace political economy as the guiding hand in British social science. Yet for us to understand the ideas that Lowe was brought to defend by Ingram, we need to turn first to the classical system of political economy and how it came to assume a position of such importance in nineteenth-century Britain.

Political economy: the British social science

'Our foreign trade, our colonial policy, our poor-laws, our fiscal system', argued J. E. Cairnes, professor of political economy at University College London, in 1873, 'each has in turn been reconstructed from the foundation upwards under the aspiration of economic ideas.' As a consequence, Cairnes went on, '[T]he population and the commerce of the country, responding to the impulse given by the new principles operating through those changes, have within a century multiplied themselves manifold.'[7] Three years later, Walter Bagehot, political commentator and editor of the *Economist*, concurred: because of political economy, 'the life of almost everyone in England...is different and better'.[8]

On account of opinions such as these, British social science during the first three-quarters of the nineteenth century generally meant one thing: political economy. Indeed, for the British, political economy was a science that meant much more than the study of economic phenomena.

The methods and ideas of political economy that captured the imagination of Victorians such as Cairnes and Bagehot were what we now know as its 'classical' form of thought. As exponents of the classical system understood it, political economy had begun with the work of the Scottish moral philosopher Adam Smith in the late eighteenth century but been given a formal analytic framework by the English political economist David Ricardo in the 1810s and 1820s.[9] In essence, this tradition of thought was rooted in a conception of political economy as an enterprise dedicated to the study of human behaviour and how the interactions between people could give rise to rules according to which public life should be governed. Thus, by building on the writings of Smith, Ricardo, and others, mid-nineteenth-century classical political economists aimed to not only understand the laws of economic interaction but also bring that knowledge to bear on the organization of politics, the state and other related fields.

Working with an analytically rigorous but highly deductive system of thought that was based on a limited number of propositions, classical political economists depicted society as a largely self-regulating entity in which wealth was distributed amongst three main groups: landowners, capitalists, and labourers. In exploring that society, political economists developed a body of key doctrines that were often inspired by a belief, indebted to utilitarian philosophers such as Jeremy Bentham and James Mill, that their conclusions could be as certain as Euclidean geometry.[10] For example, T. R. Malthus' principle of population explained the relationship between a people and its means of subsistence as a matter of natural law. Furthermore, the wages-fund doctrine, which described the relationship between wages and the economy as a whole, claimed that the earnings of labourers were paid from a fixed allocation of capital. This relationship therefore meant that workers' incomes were not only determined by the size of the available work force but also fixed in a way that made industrial action futile.[11]

The development of this *laissez-faire* system of thought took place alongside a series of important economic and political changes in the UK. On the one hand, the British economy was transformed by steam power and other new industrial processes. On the other, philosophically radical ideas began to impact on political debates, as evidenced by examples such as the Factory Acts and reform of the Poor Laws.

Embracing the logic of the market, Britain gradually replaced the mercantilism of the traditional ruling elite with policies that owed a great deal to the ideas about the benefits of free trade and enterprise that were expounded by political economists. In this respect, most triumphant of all was the repeal of the Corn Laws, a long-time *cause célèbre* of the free-trade movement and a cause to which Ricardo had devoted much of his work, in 1846, which symbolized the beginning of the mid-Victorian period that some scholars have argued was as close to a classical *laissez-faire* state as there has ever been.[12]

Whilst these changes were by no means driven by classical political economy alone, political economists were an important part of the debates through which ideas about free trade were gradually embraced. At a time when there were only a handful of chairs of political economy, at Oxford, Cambridge, and University College and Kings College London, and no dedicated economics journals in the UK, most political economists, including Ricardo, who was financier, earned their living and developed their ideas outside the academy. Consequently, as scholars such as Stefan Collini and Lawrence Goldman have shown, political economy was closely entwined with commerce and government, amongst other things, in nineteenth-century British public and intellectual life. What that meant was a free exchange of ideas between those who operated in and between those fields.[13]

Good examples of the interaction between political economy, business, and government were the Political Economy Club of London, a seldom-studied dining club founded in 1821 by a group including Ricardo, Malthus and James Mill to 'support the principles of Free Trade', and the National Association for the Promotion of the Social Sciences, which was widely known as the Social Science Association.[14] Anybody who was anybody in nineteenth-century political economy, including W. Stanley Jevons, Alfred Marshall, and John Stuart Mill, passed through the Club to dine alongside leading public figures such as the social reformer Edwin Chadwick, the Liberal Prime Minister William Gladstone, and his chancellor of the exchequer Robert Lowe. Moreover, not only were political economists, including Ricardo and John Stuart Mill, present in parliament, but the contributions of leading politicians to the Social Science Association meant it was able to impact on political discussions, as it did in the case of the Taunton Commission, which resulted in the Endowed Schools Act of 1869.[15]

One of the highest profile political debates in which the ideas of political economy featured during the nineteenth century was that of democratic reform. With the rapid growth of manufacturing towns such as

Manchester and the impact of the Chartist movement, which, despite its short-term failure, led to long-term shifts in attitudes amongst the political classes, the landscape of British politics underwent important changes during the mid-nineteenth century. In response, reformers proposed that the existing property qualification for the vote should be lowered or abolished and the electoral power of constituencies rebalanced to better reflect the new geographical distribution of people. It was Robert Lowe who emerged as one of the main enemies of these ideas, which would eventually become the Reform Act of 1867, on the grounds that the franchise was 'not a question of sentiment, of rewarding, or punishing, or elevating, but a practical matter of business and statecraft'.[16] Of the various arguments he put to his opponents, Lowe drew many from his belief that the opportunity to influence the legislative process should only be open to those who acted in accordance with the principles of political economy and therefore the best interests of the state. In this sense, a reasonable property qualification was justifiable, Lowe argued, because it gave a fair indication of who, amongst a nation of individuals born with roughly equal capacities, had worked hard and who had squandered the opportunity to improve themselves. After all, 'not to take an extreme case', he contended, '600 quarts of Beer is a fair average for every adult male in the course of the year, and, taking beer at 4*d*. a pot, the consumption of 240 quarts represents an annual outlay of 4*l*. If, therefore, persons who live in 8*l*. houses would only forego 120 quarts annually, they might at once occupy a 10*l*. house, and acquire the franchise'.[17]

For Lowe, the proposed Reform Act threatened to suffocate the principles of political economy that had played such an important part in building Britain's global economic standing and shaped its political outlook. 'The working classes', he argued, 'under the modest claim to share in electoral power, are really asking for the whole of it' and if that was allowed to happen then free trade, the heart of the mid-nineteenth-century progress in the UK, would almost certainly be abandoned.[18] 'Protection is the political economy of the poor', Lowe claimed, 'simply because they are not able to follow the chain of reasoning which demonstrates that they themselves are sure to be the victims of the waste of capital which protection implies.'[19] In this sense, 'far from believing that Democracy would aid the progress of the state', Lowe was convinced 'it would impede it' because the political economy of democratic reform was 'not that of Adam Smith'.[20]

This kind of enthusiasm for classical political economy was only possible, though, because political economists were able to take credit, via their symbiotic relationship with politics and commerce, for the economic success Britain enjoyed during the middle decades of the nineteenth century. Between 1846 and 1873, the UK led the way in an industrial transition from textiles to coal, iron, steel, and capital goods, holding a constant 40 to 46 per cent share of world exports of manufactures in the process.[21] As the more sceptical assessments of some historians suggest, it is important to keep the often wildly enthusiastic claims of the Victorians about their achievements in perspective.[22] The period was a complicated time when slumps were experienced, real wages failed to improve significantly for numerous people, and the proportion of the population claiming some form of poor relief was also more or less constant.[23] Yet the fact remained, as H. S. Foxwell, the economist Alfred Marshall's colleague at Cambridge University, argued in 1887, that political economy 'was secure in public esteem on account of the commercial prosperity which set in with the second half of the [nineteenth] century' – something that 'was popularly attributed to the policy inaugurated in 1846'.[24]

Of course, not everyone in mid-nineteenth-century Britain was unreservedly enthusiastic about the principles and methods of analysis that constituted classical political economy. The ideas that, for the most part, became the orthodoxy in the UK during the middle decades of the nineteenth century were always subject to criticism and one place where the voices of discontent were particularly loud was the British Association for the Advancement of Science (BAAS). Indeed, in the Association's social science branch, Section F, the scientific character of classical political economy was questioned regularly enough for it to become a matter of concern for a number of different groups, not least the followers of Ricardo. In fact, so serious were the debates that took place in Section F that, as we have already noted, it was threatened with closure after its meeting in 1876. For this reason, what we will be able to appreciate by shifting our attention to the BAAS are not just the debates that accompanied the rise of Ricardian classical political economy during the mid-nineteenth century but also the circumstances relating to its subsequent decline.

Social science and the British Association for the Advancement of Science

In 1876, in his presidential address to Section F of the BAAS, the Liberal MP Sir George Campbell remarked that it was the object of that branch of the Association 'to follow as far as may be a strictly scientific method of inquiry, not lapsing into the discussion of political details, but attempting to ascertain the principles on which economic results are founded'. 'It may not always be possible to draw the boundary between science and practice', he went on, 'but I am sure we shall all try as much as possible to avoid matters which involve party or personal questions, and so maintain a calm and scientific attitude in our treatment of the many subjects which come within [our] range.'[25] Whilst Campbell's comments reflected the development in Britain of a general set of concerns about political economy, they were also a response to a much longer debate about the place of political economy at the BAAS. Indeed, throughout most of the nineteenth century, Section F, which had been created in controversial circumstances a little over 40 years earlier, was often a barometer of wider opinion about the leading British social science.

As Jack Morrell and Arnold Thackray explained in their study of the founding and early years of the BAAS, the social sciences were never meant to be a part of the Association when it was created in 1831.[26] For the 'Gentlemen of Science' behind the BAAS, science was meant to be a force for unification. In this sense, the kind of normative thought about controversial and often divisive questions that political economy and its supporters stood for was exactly what the founders of the BAAS wanted to exclude from their meetings. However, and much to the Gentlemen of Science's disapproval, the calls for the BAAS to host a section devoted to questions about society and, by extension, social policy quickly became too loud to ignore. Thus, in 1833, two years after the creation of the BAAS as a whole, what was later to became known as Section F was established and began to host discussions about the application of science to issues of social and political concern.

It is often believed that the catalyst for the formation of the BAAS's statistical section was the occasion of the Association's 1833 meeting in Cambridge, which was attended by the Belgian astronomer and mathematician Adolphe Quetelet.[27] According to received views of those events, it was a group including the intellectual polymath William Whewell and mathematician Charles Babbage who instigated the move for a statistical section in the 'parliament of science' after hearing Quetelet talk about his work on crime and suicide. However, as Lawrence Goldman's work on social science in 1830s Britain has revealed, Whewell, Babbage, and their associates had been planning to agitate for the new section for some months before Quetelet's arrival.[28] In fact, far from calling for a statistical section on the spur of the moment, Whewell, Babbage, and others had a well established and highly specific agenda for statistics at the BAAS, which was based on a set of methods and aims for social science that conflicted with not only the ideals of the Gentlemen of Science but also the increasingly popular Ricardian style of political economy.

Although the BAAS had been founded with a vision inspired by Francis Bacon's 'Salomon's House', it was Whewell's philosophy of science that had almost immediately taken over at the Association.[29] Emphasizing Newtonian astronomy as the model of mature scientific enquiry, Whewell argued that it was only mature sciences that were capable of deducing general laws because only mature sciences had a great enough depth of empirical observations to make such laws possible. For this reason, it should come as no surprise that the grand deductive axioms of the relatively young Ricardian political economy were the exact opposite of how Whewell believed social phenomena should be investigated. Indeed, Whewell was convinced that the bulk of contemporary political economy to be utterly misguided. For this reason, he had been amongst the loudest of those who had called on the BAAS to create a statistical section in the early 1830s.[30]

A crucial part of Whewell's attempt to establish an alternative to Ricardian political economy at the British Association were the ideas of his associate Richard Jones, successor to Nassau Senior in the chair of political economy at Kings College, London, and to T. R. Malthus in the chair of political economy and history at the East India College. A fierce critic of the style of thought that had grown out of Ricardo's work, Jones argued that political economy needed to be a historical science that was rooted in observation and induction rather than deductive reasoning. Indeed, in the preface to *An Essay on the Distribution of Wealth, and on the Sources of Taxation*, his 1831 attack on Ricardian theories of rent, Jones had written that whilst 'Mr. Ricardo was a man of talent',

> [he] had produced a system very ingeniously combined, of purely hypothetical truths; which, however, a single comprehensive glance at the world as it actually exists, is sufficient to shew [sic] to be utterly inconsistent with the past and present condition of mankind.[31]

According to Jones and his supporters, Ricardo's methodological approach to political economy involved a sleight of hand, whereby a snapshot of a particular time and place was illegitimately presented as a universal law. Jones and Whewell were therefore in agreement about the need to base science on sound empirical foundations and, for this reason, they joined forces to promote a historically minded alternative to deductive, Ricardian political economy, first at the BAAS and then through a committee that went on to form the Statistical Society of London in 1834.[32]

Yet, despite these intentions, the BAAS statistical section was dogged from the outset by tensions between the ideals to which the group centred on Whewell and Jones hoped it would aspire, the concerns of the Gentlemen of Science that it should not compromise the harmony of the Association as a whole, and the growing popularity of the classical approach to political economy. In an effort to nullify the new section's perceived potential to compromise the disinterested authority of science that the BAAS wanted to stand for, the statistical section was therefore the subject of immediate and close scrutiny. Indeed, to avoid the introduction of opinions that might lead to clashes between politically motivated factions, the BAAS asked the statistical section to deal with facts and numbers only – a condition that other branches of the Association were never subject to. However, even those constraints failed to stop the statistical section becoming embroiled in controversy from the outset.

From its very first meetings, the BAAS's statistical section drew large audiences through sessions that were dedicated to debates on the most pressing of contemporary social questions. Although these discussions created a simmering discontent with the direction the statistical section's proceedings were taking, there was one event, the 1840 meeting of the BAAS in Glasgow, that led to those feelings boiling over. At that meeting, the gatherings of the statistical section were dominated by debates about pauperism, which enabled delegates to entertain a wide range of opinions on the political solutions for the problem. As a consequence, the statistical section's discussions of this issue were so popular they had to be moved to a larger venue. For Whewell, though, it 'was impossible to listen to the proceedings [of the statistical section]...without perceiving that they involved exactly what it was most necessary and most desired to exclude'.[33] After all, he asked, 'who would propose...an ambulatory body, composed partly of men of reputation and partly of a miscellaneous crowd, to go round year by year from town to town and at each place to discuss the most inflammatory and agitating questions of the day?'[34]

In this sense, Section F, as the statistical section became known in 1835, had immediately become a site for the type of social science, in particular a type of political economy that the section's founders had aimed to challenge. However, in following that path, Section F also posed a problem for critics who wanted it closed down. For the very same reasons it was controversial, Section F had become the best-attended branch of the BAAS and this was a consideration that could not be easily dismissed in an organization that relied on subscriptions. The Gentlemen of Science therefore had to devise some sort of compromise that would keep the section open but limit the damage it could do to the Association's wider aims and reputation. To this end, not only was William Sykes, a sympathizer with the Association's views, installed as Section F's president and given control over research grants but the section was also deliberately underrepresented on the BAAS council. Moreover, whilst the reports of the Association's meetings at first contained truncated versions of Section F's papers, potentially controversial submissions were subsequently published as abstracts only throughout the following half a century.[35]

Yet whilst Whewell, Jones, and their associates were frustrated at the BAAS, they were by no means the only people to address perceived methodological or conceptual failings in what had become the orthodox approach to political economy after Ricardo's work of the 1810s and 1820s. For example, Henry Dunning Macleod, a critic of orthodox English political economy, was one of a number of thinkers who argued the field's enthusiasm for deductive theorizing had made it a diverse patchwork of ideas that were sometimes inconsistent and often treated with greater scepticism outside the UK.[36] Furthermore, British followers of August Comte's positive philosophy were particularly critical of the normative social status political economists claimed for their work. The idea that the study of wealth alone could be a guide for social action was anathema to Comteans, such as Frederic Harrison, who argued all knowledge was building towards a great synthesis in the as yet unsubstantiated science of sociology. Indeed, Harrison regularly took his positivist attack on political economy to the pages of the periodical press, even clashing with the political economist J. E. Cairnes in the pages of the *Fortnightly Review* during the early 1870s.[37]

Nevertheless, political economy's critics faced an enormous task if they hoped to dent its public reputation during the mid-nineteenth century because, as Harrison admitted, 'that to which the cultivated public agree to look is the general diffusion of the principles of economic science'.[38] In fact, as the prominent political and moral philosopher Henry Sidgwick put it in 1887, the middle decades of the nineteenth century were 'halcyon days' for political economy in Britain:

[T]he condemnation of Political Economy by Auguste Comte was generally disregarded...I hardly think that even the eloquent diatribes of Mr Frederic Harrison induced any considerable number of readers...even to doubt the established position of economics. Nor did the elaborate attacks made by Macleod on the received doctrines succeed in attracting public attention.[39]

By 1876, though, 40 years after Whewell and Jones had unsuccessfully challenged it at the BAAS, classical political economy was 'rather dead in the public mind', according to Walter Bagehot.[40] Indeed, at the end of the 1860s the tide of British opinion had dramatically turned against political economy's classical system of thought, which kick-started a process that would see sociology enter mainstream social scientific debate. To understand why this shift happened when it did, we must consider closely the factors that led both political economists and the wider scientific and educated audience to doubt and subsequently reject classical political economy just as certainly as they had previously embraced it.

The decline of classical political economy in 1870s Britain

Addressing the Social Science Association at its 1878 meeting in Cheltenham, Bonamy Price, Drummond Professor of Political Economy at Oxford, told his audience that 'political economy at the present hour is undergoing a crisis. Both in the region of thought amongst its teachers and its students, as well as in the great world, in the practical life of mankind'.[41] Ten years later the economic historian James E. Thorold Rogers was in agreement. 'Political economy is in a bad way', he argued, 'its authority is repudiated, its conclusions are assailed, its arguments are compared to the dissertations held in Milton's Limbo, [and] its practical suggestions are conceived to be not much better than those of the philosophers in Laputa.'[42] Given the eminent position it had previously enjoyed, assessments such as these signified the extent to which political economy's reputation had declined in Britain over the course of only a decade. How and why had things gone so wrong?

According to Henry Sidgwick, it was possible to identify the exact moment when the fortunes of classical political economy began to change: in 1869 when John Stuart Mill had announced in the *Fortnightly Review* that the wages-fund theory, one of the central tenets of political economy's classical system of thought, was 'deprived of its scientific foundation and must be thrown aside'.[43] Time and again, Mill argued, it was being shown that labourers and their trades unions were capable of doing exactly what the political economists had claimed was impossible: raise pay without damaging the economy as a whole. For this reason, political economists needed to remove the wages-fund doctrine and related assumptions from their analytic toolkit. Although Mill qualified this suggestion with various warnings about the need to recognize that there was still a natural level for wages, its impact, coming from one of political economy's most widely read writers, was profound. In this sense, Sidgwick argued, Mill's recantation was an important and symbolic moment in the history of economic thought.

The decline of confidence in classical political economy, which Mill's retreat on the wages-fund doctrine represented, continued throughout the 1870s as political economists themselves questioned and then rapidly cast aside many of the doctrines and assumptions that had guided their work for the previous half century. Yet of the most important elements of the spreading spirit of criticism, almost none were new. Indeed, many were firmly rooted in the traditions of dissent against Ricardian orthodoxy that had been present in Britain since the 1820s. The Irish political economist Thomas Edward Cliffe Leslie, for example, was one of the most prominent figures in an increasingly popular historical school of thought whose intellectual lineage went back to the work of Whewell's associate Richard Jones. Echoing Jones' critique of Ricardo in 1831, Leslie argued that:

> the bane of political economy has been the haste of its students to possess themselves of a complete and symmetrical system, solving all the problems before it with mathematical certainty and exactness. The very attempt shows an entire misconception of the nature of those problems, and of the means available for their solution.[44]

Whilst these criticisms were important in casting doubt on the classical system of economic thought, they were also part of a wider set of shifts that have subsequently come to be seen as a significant moment in the development of modern economic theory. For historians of economic thought, the 1870s are associated in particular with the emergence of several ideas and works that contributed to what is now known as the 'marginal revolution', which scholars generally see as beginning in Britain with the publication of W. Stanley Jevons' *The Theory of Political Economy* in 1871 in Britain and, almost simultaneously, the work of Carl Menger in Austria and Leon Walras in Switzerland.[45] With their shared scepticism of Ricardian ideas about supply, distribution, production, and value, the political economists of the marginal revolution argued that the most pressing questions confronting their field were about consumption, which the Ricardians paid little attention to. However, the principle of marginal utility, which was at the root of an important shift in economic thought during the 1870s, had been known to political economists since the 1830s; albeit in a rudimentary form. Indeed, whilst Jevons had seen his first attempts to elucidate upon marginal utility pass largely without comment at the BAAS' 1862 meeting, his study *A Serious Fall in the Value of Gold Ascertained*, which is now considered a classic of applied economic research, sold only 74 copies on its initial publication in 1863.[46] In fact, it was only when Mill recommended one of Jevons' studies to parliament in 1866 that his intellectual stock began to rise.[47]

Although the criticisms that were levelled at political economy in the decade after Mill's recantation on the wages-fund doctrine were largely uncoordinated, there was one issue that bound them together: the perceived failure of classical theories to accurately describe the world that people were actually experiencing. As we will see shortly, what the response to this problem should be was something that political economists and the wider educated public diverged on. Nevertheless, there were reasons why concerns came to be focused on the relationship between theory and reality at the end of the 1860s. After the 'halcyon days' of the mid-nineteenth century, new economic and social circumstances had posed new questions of the doctrines and methods that had come to dominate political economy during that period. However, the responses of classical political economists were found wanting and the validity of their ideas were quickly reassessed.

Two interrelated shifts contributed most to undermining the previously optimistic view of classical political economy and its connection with economic growth. Firstly, a series of legislative changes, including the 1867 Reform Act, 1871 Trades Union Act, and 1875 Conspiracy and Protection of Property Act, transformed the British political landscape. In addition to enfranchising significant numbers of the urban population, parliament also gave workers the legal right to organize themselves into trades unions. One consequence of those changes was a growing sense that the relationships between different sections of society were far more fractious than had hitherto been thought. According to Sidgwick, this change in the general perception of the British social structure made a significant contribution to political economy's public decline. Precarious industrial relations, symbolized by a wave of strikes between 1871 and 1873, grabbed people's attention and, with the catalyst Mill provided on the wages-fund doctrine, it became commonplace to think political economists had failed totally in their efforts to comprehend the basic laws governing work, wages, and the relationship between the different social and economic classes.[48]

The second, and perhaps most important, change was in the British economy. The 'Great Victorian Boom', however that is construed, is seen to have come to an end in 1873 with the beginning of a prolonged decline in the standard indicators of fiscal progress – something Robert Giffen, one of the leading statistical economists of the age, documented in his study of an unexpected drop in prices and squeeze on profits that occurred after that date.[49] Known to economic historians before the 1930s as the 'Great Depression', the last third of the nineteenth century is often seen as a time when the UK's mid-century superiority was steadily eroded by foreign competition, in particular the rising power of German industry. Indeed, as a fall in the UK's share of world exports of manufactures from 45 per cent in the middle of the century to 31 per cent in 1900 should indicate, a shift took place in which global economic power was redistributed.[50] However, British industrial performance was by no means woeful in the closing three decades of the 1800s. In fact, whilst development was by no means as expansive as the preceding period, the quadrupling of coal exports demonstrate that Britain still performed well in some areas.[51]

Notwithstanding these more positive aspects of economic performance, the change in the overall picture of the UK's relationship to the rest of the word dealt a severe blow to previously positive assessments of the direction Britain was heading in. Indeed, such was the sense of decline when it came to discussing these issues that some commentators, including the chemist and politician Lyon Playfair, were able to use it to push their own political agendas.[52] For classical political economy, which had derived its reputation from a mutually supportive relationship with the economic and social order of the previous 50 years, these changes were bad news. As Cairnes observed in 1873, political economy's inability to adapt meant that it was 'very generally regarded as a sort of scientific rendering of *laissez-faire*', existing only to justify the economic and social form that had accompanied previous success.[53] With absolute conviction in its teachings being replaced with doubt, classical political economy's status as *the* social science in Britain was undermined accordingly.

As a consequence of this displacing of classical political economy from the centre of social scientific and political thought during the mid-1870s, an intellectual space opened up in Britain when it came to explaining how economic life was related to social phenomena. Indeed, according to Jevons, 'it [was] evident that a spirit of very active criticism [was] spreading, which [could] hardly fail to overcome in the end the prestige of the false old doctrines'. Yet, he went on, it was by no means clear 'what [was] to be put in place' of the old ideas.[54] A minority of people, including Cairnes, believed that with a great deal of effort the old orthodoxy could still be salvaged. Jevons and others believed political economy needed to be subdivided, formalized, and, in some cases, mathematized. However, for others still, the debate about classical political economy hinted at the need for an entirely different form of action: a complete overhaul of the social sciences that would involve a new subject, sociology, claiming much of the intellectual territory once occupied by classical political economy. Indeed, as we will now see, just as its rise had been fought at the BAAS in the 1830s, the decline of classical political economy in Britain and the related rise of sociology would be played out during the 1870s in Section F too.

Section F and the 'meaning of the word "scientific"'

On 31 May 1876 'a body of statesmen, economists, and statists, British, Continental, and American, such as are seldom seen together' gathered at the Pall Mall restaurant in London for the Political Economy Club's celebratory dinner in honour of the centenary of the publication of Adam Smith's *Wealth of Nations*.[55] The evening's discussion, chaired by Gladstone, was based around one question that had been proposed by his former chancellor of the exchequer, Robert Lowe. 'What are the more important results which have followed from the publication of the *Wealth of Nations*, just one hundred years ago', Lowe had asked the Political Economy Club, 'and in what principal directions do the doctrines of that book still remain to be applied?'[56] Swimming against the tide of popular opinion, Lowe offered his audience the following assessment of the state of the science: '[T]he controversies that we now have in Political Economy, although they offer a capital exercise for the logical faculties, are not of the same thrilling importance as those of earlier days; the great work has been done.'[57] Writing in the *Fortnightly Review*, Jevons was concerned by this assessment. 'One economist after another', including William T. Thornton, Cairnes, Leslie, and Macleod, had all 'protested some one or other of the articles of the old Ricardian creed.'[58] For this reason, Jevons wrote, it surely had to be accepted that 'the state of the science [is] almost chaotic'.[59]

Nowhere were the troubles that Jevons referred to more apparent than at the BAAS where, shortly after the close of its 1876 meeting in Glasgow, complaints had been made about 'the character of some of the Sectional Proceedings'. Despite the seemingly general nature of these complaints, they were in reality all about Section F. The BAAS council had therefore appointed a committee, which included Francis Galton, a writer on anthropology and travel who had recently reinvented himself as an authority on heredity and statistics, amongst its members, to 'consider...the possibility of excluding unscientific or otherwise unsuitable papers and discussions from...the Association'.[60] In so doing, the BAAS provided a focus for a new debate about the place of social science in its proceedings, which was the direct descendant of the discussions about Section F that had taken place at the Association during the 1830s and 1840s. In the late 1870s, though, the stakes had been raised by the preliminary conclusions of the BAAS' committee. With 'regard to the subjects of many papers read in [Section F], and the nature of the discussions thereon, the growth of the purely scientific branches of the Association, and to the fact that a society has been specially formed for the discussion of social and economical questions', the committee suggested that the BAAS council should 'seriously...consider whether the time has not now arrived when the section might cease to form a part of the British Association'.[61]

On account of its long-standing connection to Section F, which had been established by the BAAS committee that had helped create it during the 1830s, the Statistical Society of London (SSL) had always followed events at the BAAS closely and this new debate concerned them greatly. Indeed, at the SSL's annual meeting of 1877, the statistician and epidemiologist William Farr raised the BAAS committee's report as a serious issue, which led the SSL's council to request that their 'Secretaries...write in reply to Dr Farr' to express 'themselves strongly in favour of the Continuance of the Section'.[62] Moreover, as a sign of how strongly they felt about the matter, the SSL's council also resolved to send three of their number, including Jevons, as delegates to the upcoming BAAS meeting – a decision that led the General Secretary of the BAAS to call the SSL's 'attention to the Rule by which [it] is allowed to nominate only one'.[63] In Plymouth that year, however, the BAAS council announced that it needed to give serious thought to the criticisms that has been raised about Section F. Although a series of minor rule changes regarding the submission of papers to BAAS meetings the committee had suggested were accepted, the council asked the committee to 'report more fully the reasons which had induced them to come to this conclusion'.[64] By the end of the 1877 meeting two reports, one by Farr and another by Galton, had been presented in response to this request. Whilst the opposing arguments received by the council were ostensibly about the immediate fate of Section F, they were also concerned with a more profound and important set of differences regarding the future of social science in Britain.

The point of departure in Galton's and Farr's views was the issue that had coloured debate about Section F since its creation in the early 1830s: its popularity. For Galton the problem was that the section attracted 'much more than its share of persons of both sexes who have had no scientific training', which meant 'its discussions [were] apt to become even less scientific than they would otherwise have been'. Yet despite the fact that the dilution of Section F's already dubious scientific credentials was a problem in its own right, Galton, echoing the worries of Whewell and others in the 1830s, was also concerned by another issue: how events in Section F might impact negatively on perceptions of the Association's overall aims. With respect to this concern, Galton wanted to draw the attention of the BAAS council to the way that 'any public discredit which may be the result of [Section F's] unscientific proceedings [had] to be borne by the whole Association'.[65] For this reason, Galton argued that the BAAS could not afford to ignore what happened in Section F because the fortunes of every other section were bound up with it.

Farr's response to these charges against Section F, which had been requested by the council of the SSL in July, included a transcript of the letter from Giffen and Hammond Chubb, the secretaries of the SSL. In one respect, Giffen and Chubb had admitted in their letter, the critics were right. 'Section F is probably exposed more than any other section', they wrote, 'to the invasion of people interested in its subjects who have no scientific knowledge or training'. Because it dealt with questions bearing on everyday life, Section F understandably attracted those 'who are politicians and philanthropists, but who are not men of science'.[66] However, whilst there was no doubt that unsuitable papers and discussions had found their way into Section F, 'it could easily be shown', Farr argued, that it was 'unfortunately the case with other sections; and in all should be guarded against, as they will be, by the rules the Council has now established'.[67] Yet, Giffen and Chubb argued, it also needed to be admitted that, for all of its faults, Section F actually did a great deal of good too. Few of those who attended meetings of Section F would ever pass through any of the other branches of the Association. Thus, they went on, 'scientific men have a better opportunity [in Section F] ...than in any other [section] of communicating some notion of scientific method and its value, and of the conclusions of scientific study, to the unscientific multitude'.[68]

For Galton, though, the problem with Section F ran far deeper than a superficial concern with the numbers of people attending its meetings. Having surveyed the papers given to Section F between 1873 and 1875, he believed 'the general verdict of scientific men would be that few of the subjects treated of fall within the meaning of the word "scientific"'. 'Usage has drawn a strong distinction', he asserted, 'between knowledge in its generality and science, confining the latter in its strictest sense to precise measurements and definite laws, which lead by such exact processes of reasoning to their results, that all minds are obliged to accept the latter as true.' For this reason, whilst Section F clearly dealt with 'numerous and important matters of human knowledge', it could not be said to be a scientific body. After all, the BAAS did not consider history an appropriate subject for discussion at its meetings.[69] What confirmed Galton's suspicions that Section F failed to deal with issues governed by laws, solvable by quantification, or subject to the scientific method, was the fact that, of the papers he had surveyed, 'not a single memoir treats of the mathematical theory of statistics'. Indeed, Galton concluded, if all the unscientific papers were stripped away it would be 'impossible to continue the Section, owing to the experienced difficulty of finding suitable materials'.[70]

Of all these criticisms, it was the 'unscientific' charge that rattled Farr more than any other. Section F covered not just '*Economic Science*', which clearly dealt in facts, 'scientific determination and numerical expression', but also '*Vital Statistics*', which gave an overview of matters such as mortality and crime. The data regarding these issues were clearly subject to laws of 'so much importance', Farr argued, 'that they are observed and registered at great cost by every civilised Government in the world'.[71] Furthermore, such questions had been discussed at the BAAS since its inception by the finest of scientific minds, including Charles Babbage, Thomas Tooke, Nassau Senior, Edwin Chadwick, Jevons, and Quetelet. In this sense, Farr argued, the accusation that Section F was somehow unscientific was not one that rang true.

Giffen and Chubb, too, tried to answer this most serious of charges. Section F's critics had a vision of science that was not only narrow and restrictive, they argued, but also unable to accept that there were 'other subjects of a far more complex and difficult character than those which are the subject matter of these physical sciences'.[72] Indeed, as Karl Pearson would remark of these events some years later, Galton's definition of science was 'that of a mathematical physicist ... and rigidly applied it would exclude large regions of biology, including possibly the doctrine of evolution'.[73] 'It would therefore seem to be a degradation of the British Association', Giffen and Chubb concluded, if 'the ... whole subject of the life of man in communities[,] although there is scientific order traceable in that life, should be excluded from notice'.[74]

Over the following months the fate of Section F was occasionally the subject of discussion in the periodical press. The *Economist*, for example, published not only a supportive editorial, which argued that Section F was of the utmost importance to the study of economics in Britain, but also a lengthy, anonymous letter that was subsequently reprinted in the *Journal of the Statistical Society of London*.[75] By July 1878, with the BAAS meeting in Dublin on the horizon, the issue still loomed large, though. At the SSL they had obtained from the organizing committee an advance list of papers for Section F and, although more satisfactory submissions were expected, it appeared that 'the subjects refer[red] mostly to Social and Legislative matters, rather than to Statistics in their Scientific aspect'.[76] However, despite these concerns, the Dublin meeting did mark the closing of the Section F debate as the BAAS council declared themselves to be 'of the opinion that the existing Rules of the Association', with additional protective measures regarding the planning of the meetings, would 'afford...a sufficient guarantee for the exclusion of unscientific and unsuitable papers'.[77] Yet a simple administrative ruling could not cover up the fact that there were profound differences of opinion about the nature and methods of investigating social phenomena. Seeking to address those divisions, J. K. Ingram, the president of Section F in 1878, stepped forward to suggest that they, as well as the more general crisis in political economy, could be healed if social scientists embraced a new subject: sociology.

J. K. Ingram and the new agenda of sociology

In many ways, Galton's critique of Section F seemed to have little to do with the problems that had recently engulfed political economy in the UK. However, in addressing this issue in his presidential speech, Ingram attempted to dispel any doubts that there may have been about the connections between the two sets of difficulties. 'It is plain', he argued, 'that the low estimate of the studies of [Section F] which is entertained by some members of the Association, is no isolated phenomenon.' In fact, the spirit of criticism that Galton's report to the BAAS embodied was 'only the counterpart...of a crisis in the history of economic science'.[78] Clearly, Ingram went on, if 'the general mass of the intelligent public entertained strong convictions as to the genuinely scientific character' of political economy then there would have been no controversy about the place of Section F at the BAAS between 1876 and 1878.[79]

As he went on to explain to the audience gathered at the meeting of Section F, Ingram believed their recent troubles had arisen from failures on the part of political economists. The reason, he argued, was that the kinds of criticisms that Galton had levelled at their activities stemmed from a confusion of two distinct issues: whether 'economic facts...admit of scientific investigation, and the quite different opinion that the hitherto prevailing mode of studying those facts is unsatisfactory'.[80] According to Ingram, there was no need to make a case when it came to the question of whether one could or should apply scientific methods to economic phenomena. However, he went on, what the debate about Section F had demonstrated quite clearly was that there was a need for a new agreement about what constituted a proper scientific method and how exactly it should be brought to bear on economic and social phenomena.

Building on this theme, Ingram dedicated most of his address to a detailed and sophisticated dismantling of the, by then, widely discredited doctrines of classical political economy. In so doing, he was quite open about the sources of his own inspiration for criticizing those ideas. 'Considering the criticisms of [Comte] to have been perfectly just when he wrote them, and only requiring a certain correction now in view of the healthier tendencies apparent in several quarters since his work was published', Ingram wanted to 'dwell at some length on the several grounds of [Comte's] censures, stating and illustrating them in [his] own way'. Echoing the common concern about the growing gap between the theories of political economy and the empirical evidence to which they were supposed to relate, Ingram reduced his Comtean critique to four main interrelated issues on which he believed political economists had been misguided:

> [F]irst...the attempt to isolate the study of the facts of wealth from that of the other social phenomena; secondly...the viciously abstract character of many of the conceptions of the economists; thirdly...the abusive preponderance of deduction in their processes of research; and fourthly, the too absolute way in which their conclusions are conceived and enunciated.[81]

Whilst Ingram believed these problems could be solved easily if political economists made simple changes to the tools they used in their work, he thought the greatest concern was the status of political economy with respect to other fields. More specifically, he believed political economists needed a new framework of guiding principals to help them avoid the excesses of deduction, which had led them to present their conclusions as absolute. Indeed, he believed the BAAS could play a leading role in developing this framework. 'What appears to be the reasonable suggestion', Ingram told the BAAS, 'is that the field of [Section F] should be enlarged, so as to comprehend the whole of Sociology.'[82]

Elaborating on this suggestion, Ingram observed that what the previous two years of debate at the BAAS had shown beyond any doubt was that the kinds of mistakes both Galton and Farr had highlighted in their reports of 1877 could not be repeated. 'If we are to be associated here with the students of the other sciences', Ingram told Section F, 'it is our duty, as well as our interest, to aim at a genuinely scientific character in our work'. Whilst other societies and organizations could assume responsibility for the topics that were more a matter of politics than science, Section F's 'main object should be to assist in fixing theoretic ideas on the structure, functions, and development of society'.[83] However, this goal could only be achieved, he told the BAAS, if political economy was not treated as the social science *par excellence* but as a subject focused on the study of wealth, which was subsumed within the true science of society: sociology.

Ingram's openness about the origin of his ideas in Comte's work also revealed a significant point about how the terms of the British debate had changed by the time he addressed Section F in 1878. By his own admission, Ingram's analysis did not depart from Comte's, nor did it deviate from the criticisms Frederic Harrison had aired in the *Fortnightly Review* some ten years earlier.[84] Yet whilst the impact of such critiques on mainstream opinion had previously been negligible, the response to Ingram was immediate and positive. Indeed, for Ingram there was 'a vote of thanks [at the BAAS]...proposed by Mr Shaw Lefevre MP. Seconded by Dr Hancock...put to the meeting by the Right Hon the Lord Mayor and carried unanimously'.[85] Moreover, not only were the print media, including *The Times*, enthusiastic about Ingram's speech but it was also published as a pamphlet.[86] In fact, the address was recalled regularly throughout the final two decades of the nineteenth century not just by speakers at the BAAS but in British discussions about social science as a whole. From the NAPSS to the periodical press, an interest in the potential of a new science of society was sparked and, although not everyone endorsed the content of Ingram's positivist vision of sociology, the idea itself inspired praise in a way that it had never done before.

Welcoming Ingram's speech, Lowe, the epitome of the old political economy, advised readers of the *Nineteenth Century* to consult one of the printed copies.[87] The notion of a new social science, capable of the things Ingram claimed, was a good one, Lowe argued, but there was a problem: sociology had no methods, doctrines, laws, or teachers that could unite it as a discipline. Indeed, Lowe went on, Ingram's case for political economy to be subsumed in sociology required people to:

> admit that [such a science] exists at the present time ... that this science has already advanced so far in its construction, and produced results so clear and so satisfactory, that we are justified not only in pursuing it, but in treating it as a *scientia scientiarum* ... We are required to believe that, in a matter so infinitely various as the social state of man under every condition of climate, culture, religion, and government, this new science has already acquired such stores of knowledge, thoroughly sifted and carefully and soberly generalised ... So complete is the victory, so absolute the success, of the science of sociology, that it is able to legislate not only for itself, but for all who have come or who may come after it.[88]

In this sense, whilst Lowe was prepared to admit that political economy was experiencing problems, he thought it 'the very height of rashness and presumption to abandon what we have, before we have anything wherewith to replace it'.[89] Indeed, despite being motivated by a deep-seated, long-term attachment to classical political economy, Lowe's critique of sociology was in many important ways an incisive statement of the subject's status in Britain at the time. In a cutting turn of phrase, Lowe remarked that 'if it indeed be a science and merit one-hundredth part of what is said of it, [sociology] certainly deserves a better name than a half Greek, half Latin compound, to which it is impossible to attach any definite idea'.[90]

As was recognized both then and now, Ingram's presidential address was a significant moment in the history of British social science and for sociology in particular.[91] Delivered at a meeting of one of the most highly respected science organizations in Britain, Ingram's words had an impact of two kinds. Firstly, his critique and suggestions for the future of research into social phenomena had symbolically ended the crisis in Section F that had resulted from a new entangling of the BAAS' conception of science and the fortunes of political economy. Secondly, Ingram's speech had opened up a new debate about the possibility of a new science to fill the void that had been created by the decline of the classical form of political economy in Britain from the late 1860s onwards. However, as Lowe's response to Ingram indicated, whilst the case for sociology had been made, there was no clear and easy definition at hand of what that new science of society was, either in terms of its methods or its subject matter. Yet after 1878 a wide range of people, who were inspired by both the content of Ingram's address and the issues that had been debated in Section F, pursued new visions of social science that would see them intersect time and again with the debate about sociology. One of those people, Francis Galton, had been right in the midst of the action in Section F during the two years before Ingram's address. What we will see now, though, is how Galton's efforts to produce a social science that lived up to the ideals he had outlined in his criticisms of Section F led him to reconnect, some 25 years later, with Ingram's call for sociology.

Part II

2
Francis Galton and the Science of Eugenics

In 1877 Francis Galton (1822–1911), who had announced earlier that year the first statistical evidence of a law governing inheritance, told the committee formed to consider the future of the British Association for the Advancement of Science's Section F that 'usage has drawn a strong distinction between knowledge in its generality and science'. 'The latter' is confined, he argued, 'to precise measurements and definite laws, which lead by such exact processes of reasoning to their results, that all minds are obliged to accept the latter as true'.[1] It was therefore only possible to be scientific about society, he believed, if one proceeded according to principles and methods recognized as belonging to the natural and mathematical sciences. This chapter will show it was as a consequence of this conviction that between 1865, when he first set out his vision of society improving itself by seizing control of the processes governing heredity, and 1903, when the Sociological Society was formed, Galton developed a social, political, and scientific programme known as 'eugenics'. Indeed, what we will see is that it is only by resetting the account of Galton's work against the backdrop of the late nineteenth-century debate about social science, the subject of the previous chapter, that we can truly understand the intellectual origins of eugenics and why it came to matter not just to biology but to sociology as well.

Galton, who conceived and almost single-handedly promoted eugenics in Britain during the last four decades of the nineteenth century, has always been a historiographically complex figure. Although he briefly trained as a medical practitioner, he was neither educated in the natural sciences nor the holder of an academic post. Indeed, when compared to his professionalizing contemporaries, such as T. H. Huxley, Galton, who lived off of inherited wealth after the death of his father in 1844, has often seemed completely out of step with the trends historians are used to discussing with respect to late nineteenth-century science.[2] Yet in the generation that followed his own, Galton's work was embraced and appropriated by biologists and statisticians, such as Karl Pearson, W. F. R. Weldon, and William Bateson, who went on to play significant roles in defining the content and methodology of the natural, mathematical, and social sciences during the twentieth century. Moreover, because of the wide-ranging and creative methods he deployed in pursuit of his goals, Galton is also considered to have made contributions to the form and content of a number of other fields, including psychology.

Galton's most enduring image, though, is as the 'founder of the faith' of eugenics.[3] In this sense, his late nineteenth-century determination to prove that it is possible to manipulate and improve society through the laws of inheritance is seen as an important contributing factor to the state-sanctioned biosocial programmes, such as forced sterilization in Europe and the USA, which were implemented after his death. The consequence of seeing Galton's ideas as the preface to twentieth-century events has been that few scholars have considered his work in a way that he would have understood it himself. To be sure, following Donald MacKenzie and Ruth Schwartz Cowan, it has become commonplace to attribute Galton's views on human heredity to his socio-economic background.[4] Moreover, Galton's many biographers, including Karl Pearson, have provided us with numerous valuable insights into the man and his work.[5] The problem, however, is that the context in which Galton was working and the issues that informed his writings have largely been lost through the tendency to see him as either a lone genius or the first link in a chain. The solution to this historiographic omission, I will suggest, is to trace the development of his ideas from 1865 forwards, rather than viewing them from the perspective of where they ended up in the twentieth century.

Building on Cowan's pioneering research, this chapter takes a fresh look at how Galton put together his programme for eugenics during the late nineteenth century.[6] Beginning with his first writings on human heredity during the mid-1860s and how they were conditioned by the debate about political economy that was explored in Chapter 1, the five sections that follow move through the most significant episodes in Galton's efforts to discover the laws governing the transmission of traits from generation to generation. Taking in his early 1870s dispute with his cousin Charles Darwin about the theory of pangenesis, as well as the innovative statistical methods and data-gathering techniques that Galton developed during the 15 years after that debate, this chapter traces the trajectory of his research through to the early twentieth century, when he found a new audience amongst biologists, who saw his statistical work as an eye-opening breakthrough in the study of evolution. In so doing, the chapter restores the meaning that eugenics originally held for Galton as a social scientific, not just political, agenda and shows how his study of heredity and therefore his wider social programme only gained momentum once he had marginalized, rather than embraced, conventional biological science.

Having outlined this account, it is important to explain that the argument for the importance of social science to Galton's programme for eugenics is mainly contextual. There is no 'smoking gun' linking Galton's ideas about human heredity with the 1860s British debate about the doctrines of political economy. Nevertheless, the argument for reconsidering Galton's programme of eugenics as a biosocial alternative to the ideas about human nature and social organization that were appropriated from classical political economy by politicians such as Robert Lowe is convincing for a number of reasons. Firstly, not only does it become obvious why Galton began writing during the 1860s on the subject he would later call eugenics but the argument also explains why he played such a prominent role in the controversy over Section F at the British Association for the Advancement of Science during 1876 and 1877. Secondly, it becomes easier to understand why it was not until the end of the nineteenth century that eugenics picked up significant scientific and political momentum. Most importantly, though, recasting Galton's thought in these terms accounts for the development that his many biographers have largely ignored but on which this chapter ends: why he and his supporters came to engage with the Sociological Society at the beginning of the twentieth century.

Linking heredity with social progress

When Francis Galton first turned his attention to the subject of heredity in 1864 he was middle-aged and most widely known for writing a popular travel companion called *The Art of Travel*, which ran through five editions in his lifetime.[7] Of the reasons that had led him to explore new intellectual terrain, the most important, he recalled in his autobiography, was 'the publication in 1859 of the *Origin of Species*', which 'made a marked epoch in [his] own mental development, as it did in that of human thought generally'. His cousin Charles Darwin's theory of evolution by means of natural selection 'demolish[ed] a multitude of dogmatic barriers by a single stroke', Galton wrote, '[arousing] a spirit of rebellion against all ancient authorities whose positive and unauthenticated statements were contradicted by modern science'.[8] Yet the species-question that occupied Darwin was of no interest to Galton. Instead, it was the idea of evolution and Darwin's scientific naturalism that inspired Galton to 'pursue many inquiries which had long interested' him. Most of these new subjects of investigation were focused, Galton wrote, on 'the central topics of Heredity and the possible improvement of the Human Race'.[9] Possessing a clear sense of what an investigation of the former would uncover, he set out a research agenda in his first writings on heredity shaped around his prescriptions for the latter. Indeed, 'on re-reading these articles' at the beginning of the twentieth century, he was struck by 'their justness and comprehensiveness'.[10]

The basis for Galton's enthusiasm for and speculations about inheritance was the observation, as he put it in a two part article entitled 'Hereditary Talent and Character', which appeared in the general periodical *Macmillan's Magazine* in 1865, that 'the power of man over animal life, in producing whatever varieties of form he pleases, is enormously great', with 'the physical structure of future generations [seeming] almost as plastic as clay'. Although little was known about the biology underpinning the process, he argued such evidence strongly supported the belief there were laws governing the inheritance of physical characteristics to be discovered. It was his 'desire', though, 'to show more pointedly than ... [had] been attempted before, that mental qualities are equally under control'.[11] Thus, in these first writings on the subject, he focused on outlining a vision of humans as organisms whose physical and intellectual capacities could be manipulated in much the same way that they held those of animals to be.

Using information he had gathered from biographical dictionaries about high-profile individuals of the past four centuries, Galton outlined his case that talent and character are inherited by showing the frequency with which men who had risen through the ranks of professions such as law were related to one another. If understood and harnessed correctly, he suggested, this fact was one that had the potential to transform the world in which they lived. After all, he asked his readers, imagine what could happen if society was able to unite continuously 'in marriage those who possessed the finest and most suitable natures, mental, moral, and physical'.[12] One way society could organize itself to achieve such an aim, Galton argued, was to devise a test of 'every important quality of mind and body', which could be used as the basis for a system of competitive exams that identified the physically and intellectually most capable members of each generation. Governments could then encourage those individuals to marry by promising to 'defray the cost of maintaining and educating' their children, who would more than likely 'grow into eminent servants of the state'.[13] Within only a few generations, he claimed, the result would be 'a galaxy of genius' whose skills and talents would enrich society and drive it forwards.[14]

In his work on statistics, class, and the growth of the eugenics movement in Britain, Donald MacKenzie has argued that it is unsurprising that Galton, an independently wealthy Victorian gentlemen, who was the product of a marriage between a Darwin and a successful banker, as well as a relative of the Wedgwoods, should have thought about inheritance in this way.[15] Indeed, as Galton admitted in the autobiography he wrote towards the end of his life, his interest in the subject was stirred having first 'been immensely impressed by many obvious cases of heredity among...men who were' at Cambridge University with him during the 1840s.[16] Yet however retrospectively suitable they seem to a man of his standing, Galton's ideas about inheritance, as well as the social and political programme that accompanied them, were, as he wrote at the time, in almost complete 'contradiction to general opinion' in the middle decades of the nineteenth century.[17]

For Galton, the most objectionable aspect of contemporary ideas about human beings and their capacities was that generally associated with classical political economy, which we saw in Chapter 1 was the science that dominated the intellectual space where political, social, and scientific debate intersected. Underpinning the self-regulating view of society that political economists had developed since the late eighteenth century was a view of humans as rational agents, 'economic man', exercising roughly equal faculties in a variety of ways that helped explain social patterns, the distribution of wealth, and the workings of the economy as a whole.[18] Whilst the status of this figure – that is, whether what John Stuart Mill called *homo economicus* was a useful fiction or a depiction of reality – had always been the subject of debate, the draining of confidence in the classical system of political economy had the effect of rendering such speculations obsolete. Fiction or not, many people, including Galton, concluded that the assumptions political economists made about human behaviour were, like the wages-fund doctrine Mill had famously renounced in the *Fortnightly Review*, wrong. As he wrote in 1869 in *Hereditary Genius*, the book in which he expanded on the content of 'Hereditary Talent and Character', Galton had 'no patience with the hypothesis occasionally expressed, and often implied...that the sole agencies in creating differences between boy and boy, and man and man, are steady application and moral effort'.[19] Indeed, it was in 'the most unqualified manner' that he 'object[ed] to pretensions of natural equality' and it was towards countering such beliefs that his work on heredity was marshalled.[20]

The fact that Galton was so opposed to the then ruling ideas about humans and society also helps explain something that interpretations such as MacKenzie's do not: why Galton's programme for social improvement based on heredity first appeared when it did. As we saw in Chapter 1, the second half of the 1860s was an important time in Britain when the political, economic, and social trends that underpinned the rise of classical political economy had begun to falter, marking the beginning of the end of that science's 'halcyon days'. Whilst economic competition from overseas, particularly from the USA and Germany, was then reining in the UK's dominance of trade and production, there was also a new political agenda, embodied by the Reform Act of 1867, which had grown from the social and demographic changes brought by industrialization and manufacturing capitalism. Consequently, there was a growing sense of unease in some quarters of British society about the direction in which the country was heading and the prospects of the kind of progress witnessed in the recent past being repeated in the future – a set of concerns that dovetailed with worries about the prospects for human evolution in the wake of the publication of the *Origin*.[21] Indeed, such was the sense of decline in the UK at the time that social reformers, such as the prominent and politically active chemist Lyon Playfair, were able to use it to advance their political agendas.[22]

In this sense, as Cowan has briefly noted, Galton's emergence as a writer on heredity during the mid-1860s was bound up with wider concerns about the state of the nation.[23] More specifically, his new direction was a response, inspired by his cousin's work on evolution, to the feeling that what was going wrong in the UK was attributable to the principles of political economy that had been used to guide the country during the previous two decades; principles that had become a baffling set of predictions bearing little relation to events in the real world. On Galton's account, it was the belief in the natural equality of humans that was most to blame because, in adhering to it, the population had proven itself to fall short of meeting the challenges of modern life. 'We are living in a sort of intellectual anarchy', he wrote in 'Hereditary Talent and Character', 'for the want of master minds':

> We want abler commanders, statesmen, thinkers, inventors, and artists. The natural qualifications of our race are no greater than they used to be in semi-barbarous times, though the conditions amid

which we are born are vastly more complex than of old. The foremost minds of the present day seem to stagger and halt under an intellectual load too heavy for their powers.[24]

Thus, in the mid-1860s, Galton had outlined his belief that what little was known and could be reasonably inferred about heredity had the potential to solve the problems that were then starting to engulf political, economic, and social scientific debate. However, the vision of a society engineered by people who were in control of the processes by which they themselves had been created was in need of much more evidence to substantiate it – certainly more than the 20 pages offered in *Macmillan's Magazine*. Galton's next move was therefore to greatly expand on his writings of 1865 and try to make good his promise that there were laws of heredity waiting to be discovered.

Making sense of the data and trying to prove the case: error theory and pangenesis

Writing under the title 'Hereditary Improvement' in the general periodical *Fraser's Magazine* in 1873, Galton reminded his readers of the reasons why he believed his ongoing research into heredity to be of such importance. It was not the case, he argued, that 'philosophise as you will, men and women will continue to marry as they have hitherto done, according to their personal likings' and that, consequently, the 'prospect of improving the race of man is absurd and chimerical'. On the contrary, it was 'feasible', he argued, 'to improve the race of man by a system which shall be perfectly in accordance with the moral sense of the present time'.[25] Nevertheless, he went on, it was a challenge to create a population capable of meeting the demands of modern industrial society through the introduction of 'influences which shall counteract and overbear' the processes that were holding society back.[26] To that end, between 1865 and this restatement of his programme for social improvement in 1873, he had engaged in testing and elaborating his hypothesis about heredity in ways that both substantiated what he had done so far and shaped the direction of his project in the future.

The first major development of Galton's project came in 1869 with the publication, previously mentioned, of *Hereditary Genius* – a book already in progress when he wrote the *Macmillan's* articles.[27] At over 380 pages, *Hereditary Genius* was a massive expansion on the empirical base of the argument he had offered in 'Hereditary Talent and Character'. In the middle 13 chapters of the book, which are devoted to different professions, from English judges to wrestlers of the north country, Galton focused on men he had come to call 'eminent', each of whom was 'a leader of opinion…an originator…a man to whom the world deliberately acknowledges itself largely indebted', and the familial connections between these men.[28] According to Galton, these eminent men represented just one in 4000 people whom he believed he could show were often related to one another; a quite compelling case, he thought, for his argument about the power and importance of heredity.[29]

Galton developed his work in a second significant way when he attempted to be not just quantitative with his data but interpretative as well. His aim was to convey more precisely the meaning of eminence and to do so he had experimented with 'the far-reaching application of that extraordinarily beautiful law' known as the law of frequency of error, to which the mathematician William Spottiswoode had introduced him.[30] Originally a mathematical tool used by astronomers, the law had shown repeated attempts to take a single measurement produced figures that clustered around a central point and became less frequent at the extremities. During the middle decades of the nineteenth century, though, some writers, in particular the Belgian astronomer and mathematician, Adolphe Quetelet, had argued that the law of frequency of error could be applied more widely, including to humans.[31] However, whilst such applications were pursued with great enthusiasm, there were conceptual problems with those uses of the law that needed to be addressed. For example, it was unclear if the measurements of humans could be treated in the same way as the errors encountered by astronomers. After all, as Galton recognized, there could be no such thing as a correct height.

Despite these issues, Galton saw the law of frequency of error as a potentially significant way to understand populations, since it offered a way of explaining how men of eminence related to the rest of the population. Nevertheless, to use the law in connection with the characteristics that were his central concern, he had to find a way of measuring the talents he had claimed are inherited. The solution he hit upon was a set of exam results obtained from the Royal Military College at Sandhurst. Dividing the possible scores into classes of equal size, he produced a table with two columns: the first showing how many students would fall into each class according to error theory and the second recording the numbers that actually did (see Figure 2.1.) Although the fit was not exact, Galton thought the comparison persuasive enough to conclude that if 'everybody in England had to work up some subject and then to

Number of marks obtained by the Candidates.	Number of candidates who obtained those marks.	
	A. According to fact	B. According to theory
6,500 and above	0 ⎫	0 ⎫
5,000 to 6,500	1 ⎪	1 ⎪
5,100 to 5,800	3 ⎪	5 ⎪
4,400 to 5,100	6 ⎬ 73	8 ⎬ 72
3,700 to 4,400	11 ⎪	13 ⎪
3,000 to 3,700	22 ⎪	16 ⎪
2,300 to 3,000	22 ⎪	16 ⎪
1,600 to 2,300	8 ⎭	13 ⎭
1,100 to 1,600	Either did not venture to compete, or were plucked.	8
400 to 1,100		5
Below 400		1

Figure 2.1   Galton's comparison of Sandhurst exam scores with the results predicted by error theory.

Source: Galton 1869: 33.

pass before examiners ... their marks would be found to range, according to the law of deviation from average', with his men of eminence in the highest class and the majority of people bunched around the middle.[32] He thus extended his argument by claiming not just that physical and intellectual qualities are inherited but that they are inherited throughout the population according to the pattern predicted by error theory.

Whilst Galton's case was certainly novel and thought provoking, it was, as reviewers of *Hereditary Genius* frequently remarked, far from being as convincing as he liked to think it was.[33] To be sure, the book did not quite meet with the 'cool reception' that Karl Pearson described, but its rough edges and argumentative flaws were all noted, including the way Galton's data, which included a high number of siblings brought up in the same households, seemed to lend itself as much to an environmental interpretation as it did his own.[34] Galton had attempted a pre-emptive strike on such criticisms by adding a chapter to *Hereditary Genius* during the very final stages of the writing process, which dealt with his cousin Charles Darwin's recent 'hypothetical' theory of pangenesis – a physiological account of inheritance that Darwin had described as being 'serviceable by bringing together a multitude of facts [regarding heredity] which are at present left disconnected by any efficient cause'.[35] For Galton, pangenesis had the potential to provide an insurmountable biological defence of his argument and it was the attempt to prove this point that shaped his project in a third significant way.

Darwin had introduced his pangenesis hypothesis in 1868, in the second volume of his *The Variation of Plants and Animals under Domestication*. He had explained that there was a range of phenomena that any theory of heredity had to account for, including the apparent inheritance of acquired characters and 'reversion', whereby some offspring were thought to return to ancestral types or forms.[36] His proposal was to conceptualize inheritance in both plants and animals as a particulate process that began with each part of an organism 'throw[ing] off minute granules or atoms', called 'gemmules', over the course of its lifetime. These gemmules, which alter according to any modifications of the corresponding body parts, 'circulate freely throughout the system', he suggested, until they are drawn towards the sexual organs or buds for the purpose of reproduction.[37] Thus, on Darwin's account, offspring are created from a selection of the gemmules contained within each parent, whilst the unused gemmules lay dormant within progeny ready for transmission and potential realization in future generations.[38]

Even though it had been consciously designed to allow for the transmission of characteristics shaped by the environment, pangenesis was attractive to Galton because it gave him the sense of bringing 'all the influences that bear on heredity into a form...that is appropriate for the grasp of mathematical analysis'.[39] As he argued in *Hereditary Genius*, he believed that by using the assumptions of pangenesis there should not be:

> any serious difficulty...[standing] in the way of mathematicians, in framing a compact formula...to express the composition of organic beings in terms of their inherited and individual peculiarities...[giving us], after certain constants had been determined, the means of foretelling the average distribution of characteristics among a large multitude of offspring whose parentage was unknown'.[40]

For this reason, having seized on Darwin's somewhat vague proposition that gemmules 'circulate freely throughout the system', Galton set about testing pangenesis, hoping that by confirming its truth he would also shore up the foundations of his own argument.[41]

In a series of experiments conducted with the advice and practical assistance of Darwin and other associates, Galton, working in London between December 1869 and June 1870, extracted small amounts of blood from rabbits of one breed and colour and injected the blood into rabbits of a different kind. Galton's assumption was that if pangenesis were true then the fur colouring of the offspring of the rabbit into which blood had been injected would deviate from those that would be expected. However, in what Darwin's wife, Emma, described to a correspondent as 'a dreadful disappointment' to both Galton and Darwin, the experiments were a failure.[42] Despite some early promising signs, and even after Galton had modified his procedure and completely exchanged the blood between two rabbits, there was still no evidence, as he later put it, of 'any alteration of breed'.[43] Thus, when he went public with his findings at the Royal Society of London in March 1871, Galton announced 'the conclusion' he believed could 'not to be avoided': 'the doctrine of Pangenesis, pure and simple, as I have interpreted it, is incorrect'.[44]

Darwin, who was furious with his cousin not only because of the very public rejection of his theory but also because Galton appears not to have informed him in advance of his plans to do so, immediately composed a letter of response to *Nature* in which he took his cousin's claims apart. 'I have not said one word about the blood, or any fluid proper to any circulating system', Darwin wrote in April 1871, because:

> it is...obvious that the presence of gemmules in the blood can form no necessary part of my hypothesis; for [in the *Variation*] I refer in illustration of it to the lowest animals, such as the Protozoa, which do not possess blood or any vessels; and I refer to plants in which the fluid...cannot be considered as true blood. The fundamental laws of growth, reproduction, inheritance, etc., are so closely similar throughout the whole organic kingdom, that the means by which the gemmules (assuming for the moment their existence) are diffused through the body, would probably be the same in all beings; therefore the means can hardly be diffusion through the blood.[45]

No doubt confused by Darwin's failure to mention this objection during the course of the experiments, Galton was forced to concede his cousin's point in a letter of response, which was published in *Nature* the following week. In his defence, Galton argued Darwin's 'inappropriate use' of the words '"circulate", "freely", and "diffused"' meant the interpretation that had framed the experiments with the rabbits was the one the 'published account of Pangenesis...[was] most likely to convey to the mind of the reader'.[46] Although he could have forced such points further, Galton did not, no doubt because he realized he was involved in a debate he was never going to win. As his earlier work had made clear, Galton's interest in pangenesis was shaped by a highly specific set of issues, humans and social improvement, which made his view of the subject far narrower than Darwin's.

Despite these shortcomings, Galton continued his pursuit of a physiological account of inheritance until 1875, during which time he and Darwin even persevered with the rabbit experiments.[47] Galton did not, however, drop pangenesis but instead remodelled the theory in terms of the programme he had outlined in the mid-1860s. Whilst he maintained Darwin's notion of heredity as particulate, Galton closed the door to environmental influences by insisting that all patent and latent elements are inherited via what he called the 'stirp', which, he explained in an 1875 paper, 'A Theory of Heredity', is 'the sum-total of the…gemmules…found…in the newly fertilized ovum'. From the moment of conception, he argued, the stirp is almost totally immune from the influence of the environment and receives 'nothing further from its parents, not even from its mother, than mere nutriment'.[48]

Yet having formulated a new theory of inheritance, and one that many historians have seen as an anticipation of ideas that would subsequently become scientific orthodoxy, Galton gave little, if any, thought to the physiological mechanism of heredity again.[49] Whilst Galton's ideal scenario was for his experiments to have proven pangenesis true, the fact they did not was not a disaster. His programme for social improvement did not hang on pangenesis and, having been stung by the experience of arguing with Darwin, Galton knew he was ill-equipped to go any further with his revised theory. Galton was not a biologist and thus his most pressing concern was not with questions about the precise mechanism transmitting traits from generation to generation. Instead, he turned his focus back towards the central concern of his work during the 1860s: demonstrating *that* characteristics are inherited. In so doing, he developed his research programme in a way that enabled him to study heredity without reference to the kinds of questions about the subject that interested biologists.

Statistics, sweet peas, and the human mind

Although after the mid-1870s he had retreated from the physiological questions that concerned contemporary biologists, Galton had not given up on formulating an argument about heredity in a way that conformed to the model of scientific naturalism that had first helped inspire him to pursue the subject. He realized, though, there were no ready-made tools for him to achieve that aim. Thus, in an effort to acquire the means of legitimating his views on inheritance, Galton returned to the statistical methods with which he had experimented in *Hereditary Genius*. As we will now see, the effect of this refocusing during the mid- and late 1870s was three-fold. Firstly, as Galton's work on error theory developed, he was able to improve further on the empirical basis for his claims. Secondly, through examining that data, he uncovered the first statistical law of heredity, which vindicated his confident predictions of 1865. Finally, as his legitimating investigations gained momentum, his confidence grew, and he sought to expand his enquiries to incorporate both the measurement of physical characteristics and intellectual capacities. The consequence of these three developments was Galton's decision to temporarily cease articulating his social programme in order to focus his energies on research. When he emerged in the 1890s from 15 years of sustained investigative activity, his social programme was stronger because of it.

As Galton first explained in a letter to *Nature* in 1874, a fundamental issue with error theory was adapting it to the kinds of questions he was asking.[50] Whilst astronomers wanted to eliminate error, he was interested in the more elusive nature of the differences between people, which required a subtly different approach that could capture comparative ideas such as tall and short. He therefore suggested that the way in which he wished to apply the law of error required humans to first be scaled according to the amount or intensity they possess of the quality or attribute being investigated. Then, by picking out the individuals sitting at each quarter point in the scale, it would be possible to not only calculate what was 'technically and rather absurdly called "probable error"' but also compare two different data sets.[51] As he would later explain, the central insight of this method, which he called 'statistics by intercomparison', was the idea that it is not by measuring from the ground up that one can tell how tall or short someone is but instead by measuring from the middle of a group outwards.[52]

Yet to utilize this insight, Galton needed data. Whilst biographical dictionaries had served his empirical purposes between 1865 and 1869, they were inadequate for the case he was seeking to make in the mid-1870s. With precise human data of sufficient quantity and quality to document inheritance seeming somewhat far off, Galton's stopgap solution was to gather measurements from a suitable alternative organism: the sweet pea, which he believed to be durable and reliable enough to withstand cultivation on a large scale. As he described in a Friday Night

Discourse at the Royal Institution in 1877, he had, in 1875, 'weighed [sweet pea] seeds individually, by thousands', treating 'them as a census officer would treat a large population', and then distributed them in packets to friends and associates, including Darwin, who grew them and returned the results.[53] Consequently, Galton was able to take stock the following year of 'the more or less complete produce of ... 490 carefully weighed seeds' in which the 'only processes ... that can affect the characteristics of a sample of a population are those of Family Variability and Reversion'.[54]

To both his relief and his delight, Galton found that the law of frequency of error, which he was by then also calling the law of deviation and the normal distribution, accurately described not only the way the weights of the seeds were distributed about the population mean but also how the weights of progeny from a single family were distributed. Furthermore, the data revealed something still more significant: reversion in sweet peas, far from being arbitrary, was a tendency that 'followed the simplest possible law'.[55] As he explained to the Royal Institution audience, the mean weight of offspring from a particular family was always a third closer to the population mean than the weight of their parent. The bell-shaped curve of the normal distribution could therefore be understood, he argued, as resulting from a combination of natural variation throughout the population and, what he would later call, law-like regression to the mean in each generation.

Buoyed by his uncovering of one instance of a regularity governing inheritance, of which he had confidently predicted ten years earlier there were many, Galton's confidence in both the power of statistical analysis and the direction of his project grew. As we saw in the previous chapter, one outcome of this confidence boost was his leading role in the moves to close Section F of the BAAS, which reached its height in 1877. In light of the fact that he had made no systematic statement of his social programme for four years, and that he had made his breakthrough with the sweet peas that very year, it becomes clear why he objected so strongly to the classical political economists when he argued that they were guilty of making pronouncements on social issues without the evidence necessary to relate them to the real world. Indeed, we can see how Galton spoke at the BAAS as someone who had laboured exceptionally long and hard to acquire such evidence.

However, the confidence Galton had gained from his recent successes was also displayed at the BAAS that year in the Anthropology section, where he gave a brief glimpse of the direction he intended to take his research next.[56] Although the sweet pea research had enabled him to make firm claims about the inheritance of physical attributes, he had yet to make any advance with respect to human intellectual capacities – a controversial but important part of the argument he had put forward in the 1860s. However, the problem he had to overcome was how to go about subjecting the human mind to the kinds of quantitative methods in which he had placed so much faith. By 1877, though, he had not only hit upon several ways to go about solving the problem but also developed the momentum necessary to pursue them.

Composite photography, which he had previously used in his geographical work and discussed at length with the philosopher of evolution, Herbert Spencer, was a method Galton had designed to circumvent the challenges he faced in measuring intellectual capacities.[57] His theory was that by creating a superimposition of the images of people who possessed a particular mental quality, he could identify a physical trait that it was connected with, which would enable him to track the intellectual capacity through its physical manifestation.[58] Indeed, as he wrote in the *Journal of the Anthropological Institute of Great Britain and Ireland* in 1879, Galton thought composite photography could have other uses 'as regards enquiries into the hereditary transmission of features', such as creating representations of what the children of any particular marriage might look like.[59] To test these ideas Galton created composites using photographs of criminals convicted of murder, manslaughter, and violent robbery, which he had obtained with the help of Sir Edmund Du Cane, the Surveyor-General of Prisons. Yet despite the potential Galton saw in composite photography, which was demonstrated by the fact he continued to write about the subject throughout the 1880s, the method failed to make any significant impact on the direction or content of his research. There were, for example, all manner of problems associated with mixing images of men and women and, as he acknowledged at the Royal Institution in 1880, the final products were simply 'generic images', rather than the maps of the human mind he had believed they might be.[60]

In his other lines of enquiry, however, Galton hoped he would be able to solve, not just bypass, the problem of quantifying the mind. As he first explained to readers of the *Nineteenth Century* in 1879, he had started to keep note of the ideas and images that passed through his mind and, in the process, had been led to infer 'that [his] everyday brain work was incomparably more active... than [he] had previously any distinct conception of'.[61] Moreover, the fact that he had been able to record his ideas and images under specific conditions encouraged him to believe that they could be quantified. Coining the word 'psychometric' to describe his intentions, Galton set about refining his experiments, but he was immediately confronted with the challenge of finding a way to apply them to other people reliably. His creative solution to the problem was a questionnaire, which he designed from scratch and then distributed via friends and contacts as well as after talks he had given on his research. In addition to replies from more than 250 people, Galton also received over 150 completed forms from schoolboys, who were taught by one of his correspondents, which he explored a sample of in *Mind* in 1880.[62] Interpreting his results according to factors such as the illumination, definition, and colouring of the images reported by his respondents, Galton attempted to scale his data according to the principles of statistics by intercomparison. However, whilst he had some success in demonstrating that mental imagery conformed to the law of deviation, he only really succeeded in creating more problems. He was somewhat dismayed, for example, to find that 'the great majority of the men of science', whom he hoped would be a great source of information, 'protested that mental imagery was unknown to them'.[63] Thus, rather than solving his problems, the questionnaires threw up a whole host of unexpected issues requiring explanations.

Although these late 1870s and early 1880s investigations into mental activity would bequeath numerous tools and ideas to modern psychology, they failed to help Galton achieve what he had set out to do: bring order to the human mind and then show that its qualities are inherited.[64] Undeterred, he continued to engage, on and off, with the problem for much of the rest of his life, but as a consequence of their relative failure to achieve the results he desired, Galton returned to the general aspects of his work that seemed to persistently deliver tangible results. In the mid-1880s Galton was brought to focus once again on statistics by intercomparison and his efforts to develop a sound empirical basis for his claims by collecting measurements of physical qualities. On this occasion, however, he resolved to dispense with stopgap measures, such as the sweet pea, and instead acquire the human data that truly interested him. As we will now see, in so doing, he was, by the end of the 1880s, not only able to build upon the 1870s foundations of his work but also square it with the social programme he had set out 25 years earlier.

Anthropometrics, regression, and Natural Inheritance

In 1883, in *Inquiries into Human Faculty and Its Development*, a book in which he compiled all of his work on heredity to date, Galton signalled

that it was his intention to reconnect with the anthropocentric programme that had originally motivated that research. 'We greatly want a
brief word to express the science of improving stock', he wrote,

> which is by no means confined to questions of judicious mating, but
> which, especially in the case of man, takes cognisance of all influ-
> ences that tend in however remote a degree to give to the more suit-
> able races or strains of blood a better chance of prevailing speedily
> over the less suitable than they otherwise would have.[65]

That word, Galton argued, was 'eugenics', which he had derived from
the Greek word *'eugenes'*, meaning 'good in stock, hereditarily endowed
with noble qualities'.[66] As the pangenesis dispute with Darwin had
shown, whilst Galton was certainly concerned with universal laws of
heredity, he wished to involve himself with the generalities of the natural sciences only 'as a means of throwing light on heredity in man'. 'It
was anthropological evidence I desired', Galton declared in his presidential address to the anthropology section of the BAAS in 1885, and it
was by seizing on every opportunity to obtain it during the 1880s that
he was able finally to provide a coherent scientific framework for his
long-standing views of human and social progress.[67]

Having received a quite limited response to his mid-1870s calls for
schools and other public institutions to supply him with the kinds of
data that would be useful for his work, Galton tried two more proactive approaches in the 1880s, both of which offered incentives for people to cooperate with his enquiries.[68] The first, which he announced
in a letter to *The Times* in January 1884, resonated with his general
desire 'to encourage a habit of preserving family records to enable parents to appreciate the various hereditary influences converging upon
their children'.[69] To catalyse this process, Galton announced that he
was offering a total of £500 in prizes, approximately £40,000 in early
twenty-first century money, to those who could provide such information by filling out and returning to him by mid-May a questionnaire he
had recently published, entitled *Record of Family Faculties*, which asked
for details about family members, such as heights, weights, and health,
covering four generations.[70]

As the deadline for the questionnaires expired, Galton unveiled his
second new approach at the International Health Exhibition in the
Royal Horticultural Society's gardens in South Kensington, London,
where he attempted to make real a vision of 'anthropometric labora-
tories' in which people would be 'weighed and measured, and rightly

photographed, and have each of their bodily faculties tested, by the best methods known to modern science'.[71] In an exhibition space approximately six feet wide and thirty-six feet long, which he partially obscured with trellis-work, Galton set up a table on which he positioned a set of instruments that measured things such as height, weight, and reaction times.[72] Having paid a fee of three pence to enter this area, people were led through the various tests by a superintendent he had employed to run the anthropometric laboratory. Then, at the end of the process, participants were given a card listing their scores, of which Galton kept an anonymous copy.[73]

Between them, the *Record of Family Faculties* and the anthropometric laboratory supplied Galton with the large body of human data he had craved for since the 1860s. By the closing date for the *Record* he had 'upwards of 150 good records of different families', many of which were 'concise, full of information, and offering numerous opportunities of verification', providing him, he estimated, with references to over 5,000 people.[74] Yet the anthropometric laboratory furnished Galton with an even greater volume of information as a total of 9,337 people passed through it.[75] Indeed, so successful was the laboratory that when the International Health Exhibition closed in 1885 he moved the laboratory to a number of other locations, eventually settling in the Science Galleries of the South Kensington Museum, where he collected a further 3,678 sets of measurements by 1892.[76]

In many ways, this mine of information enabled Galton to expand on the work he had done earlier with sweet peas. However, to view what he did with this new data as a simple extension of his earlier researches would be to overlook the way his latest successes enabled him in 1889 to write *Natural Inheritance*, his most significant book since *Hereditary Genius*, which totally transformed the fortunes of his wider project. As Galton announced on the occasion of his presidential address to the anthropological section of the BAAS in 1885, the consequence of the new data had been to 'place beyond doubt the existence of a simple and far-reaching law that governs hereditary transmission'.[77] By examining '930 adult children and their respective parentages', he had been able to demonstrate that the normal distribution applied to humans in the way it did other organisms and data.[78] Moreover, he was able to show that the regression to the mean he had observed in sweet peas was true of human parents and their offspring too, which confirmed that like does not produce like; it produces something a third more mediocre than itself. Whilst this fact legislated 'heavily against the full hereditary transmission of any rare and valuable gift', he told his audience, the law was reassuringly 'even-handed' because 'it levies the same heavy succession-tax on the transmission of badness as well as of goodness'.[79]

Although this new breakthrough owed a great deal to Galton's empirical endeavours, it was one he had made with the aid of fresh statistical innovations. First amongst the new methods unveiled in his BAAS address was what he called the 'mid-parent', which dealt with the problem posed by the fact humans, unlike sweet peas, have two parents whose exact contribution to their offspring was then still unknown.[80] Having first ensured that all of his measurements were on the same comparative statistical scale, which he did in the case of height, for example, by multiplying the heights of women by 1.08, the factor he had calculated as accounting for the general difference in stature of men and women, Galton then worked out the average of each set of parents in his data. Although he had created these mid-parents more through a sense of hope than expectation, he found they were an exact solution to the two-parent problem. Indeed, as he told the BAAS, when he examined an attribute such as height he had discovered 'the stature of...children depends [so] closely on the average stature of the two parents' that it 'may be considered in practice as having nothing to do with their individual heights'.[81]

Furthermore, Galton also announced at the BAAS a more general development, which was both so technical in its nature and of such importance that he had sought independent confirmation of his conclusions. Whilst tabulating his new data, with classes of mid-parents against classes of offspring, he had noticed a curious phenomenon. When he drew a line through the squares on his table that contained the same number, he produced a series of concentric ellipses.[82] In Galton's view, this elegant and unexpected result indicated that the law of regression was far from being the only statistical relationship in his data. More specifically, what he thought he could see was that it was not just offspring that regressed on mid-parents but mid-parents that regressed on offspring too.[83] Convinced this fact would mean that the variables in his data were locked in a series of mutually dependent relationships, he had 'phrased the problem in abstract terms...disentangled from all reference to heredity' and sent it, on the recommendation of a friend, to a mathematician – J. Hamilton Dickson of Peterhouse College, Cambridge. As Galton told the BAAS, he had never 'felt such a glow of loyalty and respect towards the sovereignty and magnificent sway of mathematical analysis' as when he received Dickson's answer because it confirmed 'by purely mathematical reasoning, [his] various and laborious statistical conclusions with far more minuteness than [he] had dared to hope'.[84] Indeed, what Dickson had succeeded in showing was 'that the law of error holds throughout the investigation with sufficient precision to be of real service, and that the various results of [Galton's] statistics are not casual determinations, but strictly interdependent'.[85]

With his 20-year use of error theory vindicated, Galton answered explicitly and with great clarity one of the most fundamental questions raised by his research. 'How is it', he asked,

> that in each successive generation there proves to be the same number of men per thousand, who range between any limits of stature we please to specify, although the tall men are rarely descended from equally tall parents, or the short men from equally short?[86]

'The answer', he told the BAAS, was deceptively simple: 'the process comprises two opposite sets of actions, one concentrative and the other dispersive, and of such a character that they necessarily neutralize one another, and fall into a state of stable equilibrium'.[87] Because tall men married short women as often as short men married tall women, mid-parents across the population as a whole were effectively created through chance. There is therefore in each generation a balanced set of conflicting forces, with variation pulling offspring outwards towards the extremes and regression pulling them inwards towards the mean, which results in the elegant bell-shaped distribution predicted by the law of frequency of error.

Having announced these developments to the BAAS in 1885, Galton set about publishing papers that not only made his empirical evidence more widely available but also indicated how his methodological insights might apply to attributes other than height.[88] He realized, though, that this scattergun approach of presentations and article-length publications limited the wider impact his work could make. Galton now wanted to paint on a broad canvas and so his next move was to bring everything together in a single book that could be con-sidered the intellectual successor to *Hereditary Genius*. However, unlike *Hereditary Genius*, *Natural Inheritance*, which was published in 1889 and immediately recognized by the aspiring Scottish man of science Patrick Geddes as 'essentially a republication of a number of [Galton's] separate papers more or less statistical in treatment', did not carry an explicit statement of Galton's wider commitments.[89] Instead, Galton concen-trated on bringing together his conclusions about heredity and the methods he had used to find them during the previous two decades. Yet anyone familiar with his work could not have missed his veiled

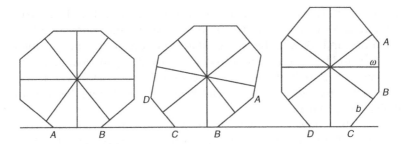

Figure 2.2   Galton's polygon of organic stability

attempt to relate his nuanced, statistical understanding of heredity to the bigger ideas that had inspired him to study the subject in the first place.

At first glance, the rigid stability of populations that Galton had uncovered, though it had precedents in his earlier work, looked out of step with the social scientific optimism of his general programme. Indeed, as he had remarked in 1883, it would seem that eugenics was a task comparable to that of Sisyphus.[90] However, by rejecting his cousin Darwin's account of evolution as a slow process proceeding by tiny, accumulative steps, Galton had developed an alternative view of evolution that reconciled organic stability with his hopes for social change.[91] As he explained in *Natural Inheritance*, Galton had come to see an evolving population or group as being much like the polygons shown in Figure 2.2: stable around a central point but capable of so being on any one of their many surfaces. In this sense, he argued that although 'a long established race habitually breeds true to its kind...every now and then the offspring of these deviations do not tend to revert, but possess some small stability of their own'.[92] In fact, these non-reverting offspring, which are known as 'sports' and occupy the far ends of the normal distribution, are so distinct they are 'capable of becoming the origin of a new race with very little assistance on the part of natural selection'.[93]

For these reasons, evolutionary progress comes about, Galton asserted, not through small incremental steps but relatively sudden shifts that were akin to a polygon being pushed with such strength that it toppled over on to another of its sides. If people felt empowered to identify and foster these sports in society, the stability he had uncovered was therefore not an enemy of social progress but an ally with which to work. As he had put it in 1883 in the conclusion to *Inquiries into Human*

Faculty and its Development, Galton wanted people to look at themselves as 'more of a freeman, with the power of shaping the course of future humanity, and...less as the subject of a despotic government, in which case it would be his chief merit to depend wholly upon what had been regulated for him, and to render abject obedience'.[94] However, it was clear that a great deal of research needed to be done before that state of affairs could be realized.

As we shall now see, much of this research was not actually conducted by Galton. Instead, it was biologists whom he had avoided since his dispute with Darwin over pangenesis during the early 1870s who stepped forward to provide Galton with even more evidence to support his vision of social reform. Indeed, despite the fact that his work was appropriated for often conflicting purposes, and then developed in ways he did not fully understand, Galton managed to establish a significant intellectual connection between himself and the late nineteenth- and early twentieth-century biological community through a number of individuals who had come to revere his work as an eye-opening, methodological triumph for the study of evolution. For Galton, the consequence of respectable men of science taking his work seriously was a strong legitimating platform in the natural sciences that finally enabled him at the beginning of the twentieth century, over three decades after he had started work on the subject, to launch eugenics as a social scientific programme for reform.

Continuity and discontinuity amongst Galton's supporters in biology

In common with all of Galton's enquiries after the pangenesis dispute, *Natural Inheritance* featured little that could be classed as 'biological' when compared with the kinds of approaches to evolution and heredity that were utilized across the biological sciences. Indeed, as Peter Bowler has argued, Galton only began to make progress in his work once he had divorced his enquiries into heredity from the physiological questions about inheritance that mattered to biologists.[95] However, as has been documented extensively, that situation changed when various aspects of his work were embraced and approved by two groups, which came to be known as the biometricians and the Mendelians after the 'rediscovery' of Gregor Mendel's work on plant hybridization in 1900.[96] That the late nineteenth- and early twentieth-century dispute between these two seemingly incommensurable groups was frequently bad-tempered did not matter to Galton, though. What was important to him was that much of the agenda in an important area of the biological sciences was being set by scientists who saw themselves as continuing a project he had started in *Natural Inheritance*. As a consequence, when Galton came to renew his calls for a social programme called eugenics, he did so with the knowledge that many of his most important ideas had taken root in biology.

In general terms, the biometricians, who were led by the statistician Karl Pearson and the biologist W. F. R. Weldon, and the Mendelians, whose most prominent member was William Bateson, were inspired by two different aspects of Galton's work. For the biometricians, it was Galton's statistical approach that was most important because they believed it to have the potential to revolutionize the way biologists studied and thought about evolution. As Weldon, who held the Jodrell chair of zoology at University College London, put it in 1894, the biometricians were convinced that 'the questions raised by the Darwinian hypothesis are purely statistical, and the statistical method is the only one at present obvious by which that hypothesis can be checked'.[97] However, they diverged from Galton when it came to the conclusions they drew from facts such as the stability of organic populations and his claim to have refuted Darwin's general account of evolution. Suggesting that Galton had misunderstood the meaning and effect of the regularities he had uncovered, the biometricians argued that evolution did proceed by minute steps and the aspects of his work that Pearson and Weldon appropriated, such as 'Galton's Law of Ancestral Heredity', which expressed the geometric contribution of ancestors to progeny, reflected this fact.[98] However, for Bateson, a fellow of St. Johns College, Cambridge, and other Mendelians, Galton's statistics were not as important as his argument for seeing evolutionary change as discontinuous; that is, as being like a polygon. Indeed, after a series of empirical studies during the late 1880s, Bateson had become convinced that the use of statistics to uncover evolutionary trends at the macro level had to be tempered by the specialist, observational skill of the trained biologist.[99] The effect of eschewing the restraints of purely quantitative approaches was, Bateson argued, to reveal a great deal of empirical evidence that supported the importance of sports in evolutionary change.[100]

Galton's relationship to these two groups throughout the last decade of the nineteenth century and first decade of the twentieth was complex. On the one hand, he developed strong links with the biometricians almost immediately after the publication of *Natural Inheritance*. As Pearson developed Galton's statistical tools way beyond his powers of mathematical comprehension, Weldon produced empirical studies, most notably on crabs, in which he used statistics to solve practical biological questions, such as how to tell closely related species apart.[101] Spurred on by the results the approach was yielding, Galton and Weldon helped found the 'Committee for Conducting Statistical Inquiries into Measurable Characteristics of Plants and Animals' at the Royal Society of London in 1894 as a vehicle for their ideas. Yet Galton was just as keen on Bateson's work and, at the same time as he was founding the committee at the Royal Society, Galton was praising Bateson's arguments for the importance of discontinuity in evolution.[102] Galton's attempt to resolve the problems raised by this dual allegiance was to bring both sides together, as he did at the Royal Society in 1897, when he opened up his committee to Bateson and other non-biometricians, renaming it the 'Evolution (Plants and Animals) Committee'.[103] However, the well-documented effect of such manoeuvring was the intellectual alienation of the two groups from one another, a fact that was symbolized by Weldon and Pearson's departure from the Royal Society committee to found the journal *Biometrika* in 1901, and a level of personal animosity that was only brought to an end by Weldon's untimely death in 1906.

Given his obvious sympathies towards the arguments offered by both the biometricians and the Mendelians, Galton's work seems to have had far from the effect one might expect him to have desired. However, whilst such divisions might not have been ideal, they were far from being a major problem for him. By reading back from the twentieth and twenty-first centuries with the knowledge of the impact Galton had on late nineteenth- and early twentieth-century science, it is easy to see him as having worked with the biological sciences in mind all along. Yet Galton had never pursued the study of heredity with the intention of transforming biology. In fact, he had tried to do without conventional biology after his experience of the pangenesis dispute during the early 1870s. The sudden enthusiasm for his work in the biological community did not therefore represent the end point for him; rather, it was a means towards the end of realizing the aims he had first outlined in 1865. As Pearson wrote of Galton's activities during the 1890s and early 1900s, 'a very fundamental characteristic of Galton's mind was his desire that our progressive knowledge of natural law should at once be turned to practical service in attempts to elevate the race of man'. Galton therefore could not 'think of the doctrine of evolution merely as a contribution to academic biology; for him the type of "sport" of greatest interest and value was that embracing the human moral or intellectual "sports" and he desired at once to know how we might perpetuate for the service of mankind such supermen as might appear'.[104]

On account of the work that he had done since the 1860s and the support it had received from natural scientists from the 1890s onwards, Galton felt confident at the beginning of the twentieth century that the general thrust of what he had written in his articles for *Macmillan's Magazine* and then in *Hereditary Genius* had been proven right: there were regularities governing inheritance and it was possible to investigate the subject with scientific tools. Thus, it seemed plausible when Galton claimed, as he did in 1891 in a presidential address to the division of demography at the International Congress of Hygiene and Demography, that 'the improvement of the natural gifts of future generations is largely, though indirectly, under control. We might not be able to originate [in evolution] but we can guide'.[105] In this sense, by the first decade of the twentieth century, eugenics had become a biologically rooted social science, despite the fact Galton had developed it without reference to the concerns of biologists themselves. For this reason, it should be no surprise that when the opportunity to promote eugenics presented itself to Galton in 1904 at the recently created Sociological Society, he did so with Pearson and Weldon in the audience and a supportive written communication from Bateson. However, as we will now see, there was more than one way to connect the study of society with the study of biology in the late nineteenth and early twentieth centuries. Indeed, for Patrick Geddes, a former student of 'Darwin's Bulldog', T. H. Huxley, being scientific about society meant bringing the knowledge that biologists had of the evolutionary process, as well as the methods they had used to acquire it, to bear on the study of how people both shaped and were shaped by the places in which they live.

3
Patrick Geddes' Biosocial Science of Civics

Reviewing Francis Galton's *Natural Inheritance* for the *Scottish Leader* newspaper in 1890, the Scottish biologist Patrick Geddes (1854–1932) observed that when Galton first published on the subject of heredity during the 1860s it had been a time when the 'Political Economy Club was wont to dine cheerily, thinking their theory of individualism was "approximately perfect"'. However, since then, Geddes explained, the Political Economy Club's 'pens [had become] rust' and all had 'been painfully convinced that we are…members of one another'. It was therefore no surprise, he argued, that Galton's approach to heredity had changed. 'Formerly [Galton] was interested', Geddes wrote, 'with the conscious pride of an intellectual patrician, himself sprung of the mighty races of Darwin and Wedgwood, in compiling a sort of spiritual peerage'. In the late 1880s, though, Galton had come to insist, Geddes continued, 'not only upon the fraternity but that it be a large one'.[1] This chapter throws light on Geddes' comments about Galton's eugenic research by reconnecting with the position he had staked out in the late nineteenth-century debate about evolution. In so doing, we will see how Geddes, inspired by work he had carried out during the late 1870s and early 1880s, put together a programme called 'civics' that was a serious biosocial alternative to Galton's eugenics in the debates that subsequently took place at the Sociological Society in the early twentieth century.

When he reviewed *Natural Inheritance* in 1890, Geddes was primarily known as a biologist and a former protégé of T. H. Huxley, the archetypal Victorian man of science. However, in the late twentieth and early twenty-first centuries, Geddes has come to be recognized more for his work in other fields. A key figure in the early twentieth-century British debates about sociology, who not only went on to be professor of sociology and civics at the University of Bombay from 1919 to 1924 but also wrote several important works on town planning, Geddes is now most often seen as a social scientist.[2] Yet on account of his earlier identity as a biologist – one who did tightly focused experimental work in a laboratory setting – Geddes' career trajectory has always been an interpretative challenge. To be sure, scholars including Helen Meller and Volker Welter have always recognized that Geddes' twentieth-century writings on cities were indebted to what he had learned from the discipline in which he first trained.[3] The problem, though, is that in making those connections, few scholars have ever looked closely at Geddes' early work as a biologist. As a consequence, the suggestion that it was the idea of evolution that held his wide-ranging pursuits together has always glossed over complicated but important issues that we have to understand if we are to grasp what evolution meant to him. Indeed, because scholars have little sense of where Geddes should be positioned in the context of the late nineteenth-century biological debates that shaped his view of evolution, we have a weak grasp on how and why he moved so freely between the sciences of life and society.[4]

These interpretative difficulties have been highlighted recently by the clash between Steve Fuller, on the one hand, and Maggie Studholme, John Scott, and Christopher Husbands, on the other, over how Geddes related biology and society in his work. Responding to Studholme's positioning of Geddes in the tradition of environmental sociology, as well as Scott and Husband's efforts to recover the life and work of Geddes' friend and collaborator, Victor Branford, Fuller has argued that our understanding of Geddes has been compromised by the needs of late twentieth- and early twenty-first century political correctness.[5] On Fuller's account, Geddes and Branford wanted to connect sociology and biology in such a way that, had Geddes been given an institutional footing in the early twentieth century, he might have associated himself with groups that his current admirers, many of whom regard Geddes as a hero of the environmentally conscious political leftwing, would be deeply uncomfortable with.[6] For their part, Studholme, Scott, and Husbands have rejected Fuller's counterfactual claim that Geddes' regionalist approach to sociology and town planning might have led him to approve of parts of the Nazi party's programme. However, what serves as the starting point for these widely divergent assessments of Geddes' later career, which will be considered closely in the conclusion of this book, is the ambiguity of Geddes' biology and how he related it to his ideas about society.

This chapter attempts to resolve these difficulties through a new account of how Geddes put together the biosocial programme that, by the first decade of the twentieth century, he had come to call 'civics'. Charting his career from the mid-1870s, when he was a student of Huxley's, through to the eve of his appearances before the Sociological Society, this chapter argues that in moving between the biological and the social sciences, Geddes balanced two different demands. On the one hand, Geddes understood sociology to be a genuinely *social* science, which was independent from other sciences and had its own distinctive subject matter. Nevertheless, he also believed sociology had to take into account its biological basis and, in this sense, he also wanted it to be rooted in the methods and theories of the kind of biology he had practiced early on in his career. Moreover, whilst these concerns were related to a number of Geddes' late nineteenth-century experiences, this chapter shows there were two encounters in particular that helped guide his attempt to balance the biological and social sides of the civics equation: his contact with the writings of Herbert Spencer and his response to the late 1870s debate about the methods, aims, and scope of the social sciences, which was explored in Chapter 1.

As we will see through a close examination of the experiments Geddes conducted in France and Italy during the late 1870s and early 1880s, his ideas about evolution, the platform for all of his subsequent work, were profoundly indebted to his appropriation of Spencer's evolutionary philosophy, which Geddes had first read when he was a student of Huxley's in the mid-1870s. Motivated by a Spencerian view of evolution and a belief that his training as a biologist could help realize the goals J. K. Ingram had outlined at the British Association for the Advancement of Science in 1878, Geddes went on during the 1880s to create a genuinely biosocial science that urged society to confront the biological sources of its identity – an argument that was made most clear in 1889 in his first book, *The Evolution of Sex*. As we will see finally, Geddes then utilized a range of other sources during the 1890s, including the ideas of the French social surveyor Frédéric Le Play and the efforts of social reformers in London, to turn his theoretical framework into a comprehensive programme that was as much about action as it was thought. However, so that we can see how it not only permeated Geddes' scientific practices but also shaped how he related the social and biological worlds, we must first recall the content and aims of Spencer's evolutionary philosophy.

Cooperation between parts: Herbert Spencer's evolutionary philosophy

Despite the important work by Robert M. Young, Robert J. Richards, and others, Herbert Spencer (1820–1903) has frequently been seen as a writer who openly used Darwin's theory of natural selection to justify a bourgeois *laissez-faire* social theory that we now call social Darwinism.[7] Indeed, when evaluating Spencer's status as a scientific thinker, many commentators have taken their lead from Ernst Mayr, who reassured us that it is 'quite justifiable to ignore Spencer totally in a history of biological ideas because his positive contributions were nil'.[8] Recently, though, a number of scholars, including Michael Ruse, in an important study of the population geneticist Sewall Wright, and Gregory Radick, in his work on late nineteenth- and early twentieth-century anthropologists and sociologists, such as A. C. Haddon, W. H. R. Rivers, and L. T. Hobhouse, have challenged this dismissive consensus.[9] Indeed, allied with the work of scholars such as Thomas Dixon and Naomi Beck, who have recaptured not just the rich content of Spencer's writings but also their diverse readership, it has started to become clear not just how narrow traditional understandings of Spencer have been but also why it was more than just free-market capitalists who were interested in his work during the late nineteenth and early twentieth centuries.[10]

As he first began to explain in the early 1850s, Spencer believed that evolution is a universal phenomenon, the effects of which are observable in a single, progressive trend permeating every aspect of existence.[11] On his account, whether one is referring to the heavens, species, or human society, evolution is the general process of the simple and homogenous being transformed into the ever more complex and heterogeneous. Thus, in his defining collection of work, the Synthetic Philosophy, which was funded by the subscriptions of readers and issued in five parts, building from philosophical first principles, through biology, psychology, sociology, and ethics, over the course of 44 years, Spencer spelled out in minute detail how evolution had produced the world in which he and his contemporaries found themselves. More than that, though, his writings were a heady mix of the descriptive and the normative in which the future was seen to depend on the fortunes of evolution in the present.

In *First Principles*, his system's foundation stone, Spencer explained how everything subsumed within the macro-trend of evolution was the result of a balance of tangible forces within it. 'The processes thus everywhere in antagonism, and everywhere gaining now a temporary and now a more or less permanent triumph the one over the other', he wrote, 'we call Evolution and Dissolution. Evolution under its simplest and most general aspect is the integration of matter and concomitant dissipation of motion; while Dissolution is the absorption of motion and concomitant disintegration of matter'.[12] Organic evolution, as he went on to explain in *The Principles of Biology*, is therefore a process that conforms to the general pattern of development, but in a manner facilitated by a conflict of forces both within organisms themselves and between organisms and their environment.[13] The Spencerian natural world is therefore a place of constant flux, where there is, in the standard phrase, a 'dynamic equilibrium'. Thanks to ceaseless change, organisms are constantly forced to confront new circumstances and, according to Spencer, it is their equilibration to those conditions that defines organic evolution.

Contrary to received views, natural selection, or 'survival of the fittest' as he sometimes termed it, did not serve as the sole or even primary explanatory mechanism in Spencer's account of development. Whilst he did include it in his writings after the publication of *On the Origin of Species* in 1859, Spencer always made clear that he did not believe natural selection to be an all-powerful force.[14] Indeed, 'on critically examining the evidence', he wrote in 1887, 'we shall find reason to think that [natural selection] by no means explains all that has to be explained'.[15] In Spencer's view, there are two possible outcomes for organisms of the shifting balance of forces that they confront – death or a new equilibrium – and natural selection is simply the negative process associated with the removal of those unable to adjust to change.[16] However, positive, creative adaptation, which he saw as the response of most organisms to the challenge of new conditions, has to happen by means other than natural selection. Thus, on Spencer's account, organisms evolve by Lamarckian means, passing on the adaptations that they acquire during their own lifetimes, and, as a consequence, both individual organisms and the species of which they are a part can be seen as following the path towards heterogeneity.

The outcome of these processes, Spencer explained in *The Principles of Biology*, is not a world red in tooth and claw. On the contrary, the destructive mode of competition that characterizes lower forms of life should gradually disappear, he argued, as highly specialized organisms tend naturally to cooperate. Spencer's reasoning on this point was based on his belief that the result of greater complexity is a higher level of integration and mutual dependence. After all, he asserted, the more differentiated something becomes, the more the functioning of the whole depends on the cooperation of the parts. 'Changes numerical, social, organic, must, by their mutual influences', he wrote, 'work increasingly towards a state of harmony…and this highest conceivable result must be wrought out by that same universal process which the simplest inorganic action illustrates.'[17] It was in this sense Spencer's perhaps unexpected conclusion, symbolized in 1892 by his issuing of *The Principles of Ethics* as the final part of the Synthetic Philosophy, was that evolution is a process with an ethical end. The belief guiding his work was therefore that evolution, if it is allowed to take its course, will continue to facilitate the emergence of behaviour that is both widely beneficial and of a higher type.

Throughout the late nineteenth century, Spencer endeavoured to trace these kinds of changes through all manner of natural and social phenomena – a move that frequently drew him into intellectual conflict. With T. H. Huxley, for example, Spencer debated how best to understand the relationship between the different parts of a living being. Whilst Spencer, following the logic of his evolutionary system, saw organisms as unions of cooperating components, Huxley held a diametrically opposed view, with organisms as collections of parts dominated by either the brain or will.[18] Given the entanglement in Spencer's thought of the natural and social worlds, arguments like this had political implications. However, most of those implications do not fit well with received views of Spencer as the arch-apologist for social Darwinism and Huxley as the rational agnostic enemy of religious authority over science. For example, it was Spencer, not Huxley, who was known to contemporaries as a pacifist and a vehement critic of imperialism, and especially of the claim that empires were somehow a beneficial and civilizing force for subjected indigenous peoples.[19] Indeed, in the 1870s, when Spencer was reaching the height of his fame, his writings were permeated by a concern with altruism, and his sociology in particular with how the gradual disappearance of the struggle for existence translated into a normative distinction between lower military and higher industrial societies.[20] With these dimensions of the Spencerian system in view, it becomes clearer why his audience comprised large numbers of people who were of profoundly left-leaning convictions.[21]

Of course, myth-busting reconstructions of the Synthetic Philosophy must always be tempered by the recognition that, particularly in the final two decades of the nineteenth century, Spencer really did propound many of the ideas traditionally attributed to him. It is important, though, to see that there was more than one message to be taken even from late-period Spencer. Unrestrained capitalism was neither the most natural nor the most desirable state of society in Spencer's philosophical system, and his belief that governments, no matter how good their intentions, are better off staying out of the affairs of individuals was a reflection of his almost utopian view of evolution and its progressive potential. Yet, since he first developed those arguments, our political sensibilities, including where we draw the line between left and right and nature and society, have shifted in such a way that making sense of Spencer has proven to be a difficult task. As someone who is frequently celebrated as a proto-environmentalist, Patrick Geddes, like many of his generation, has seldom been connected in any meaningful way with the thinker who coined the phrase 'survival of the fittest'.[22] However, as we will now see, Geddes actually built his multi-dimensional career on Spencerian foundations.

Patrick Geddes' theory of 'reciprocal accommodation'

In 1876, after a short period of study at Cambridge University with the leading British embryologists Francis Balfour and Michael Foster, Geddes, then 22 years old, returned to the Science Schools, the South Kensington branch of the School of Mines in London, where he was employed as a demonstrator by his scientific patron and former teacher, T. H. Huxley. Whilst recounting what he had done in Cambridge, Geddes mentioned to Huxley that he had read Spencer's *The Principles of Biology*. Despite the fact they were good friends, Huxley and Spencer, as we have already noted, disagreed profoundly on many of the most pressing questions of contemporary debate about evolution, not least whether it was possible to relate the natural sciences to social and ethical processes.[23] Huxley's response to his young student's news was therefore to tell him that he would 'have done far better to spend all [his] time on embryology!'[24] Having abandoned a botany degree at the University of Edinburgh in 1874 after only a week just so that he could study under Huxley, Geddes held his teacher in the highest regard. Indeed, towards the end of his life, Geddes continued to express his admiration for Huxley as 'anatomist and as man'.[25] Yet, after over a year spent in his laboratory in the mid-1870s, Geddes had grown disillusioned with Huxley's determination to keep his enthusiasm for evolution within the bounds of anatomical analysis and the biological sciences as a whole.[26] Thus, far from being put off by Huxley's forthright dismissal, Geddes, who had turned to Spencer in an effort to fill the spaces Huxley had left blank, immediately reread *The Principles of Biology* to find out more.[27]

As the research Geddes would soon produce made clear, that decision was one that had profound consequences for his view of the evolutionary process as a whole.

Although they were diverging on significant intellectual matters, Geddes' enthusiasm for biology and his technical skill around the laboratory persuaded Huxley to continue to extend to him his considerable powers of patronage throughout the late 1870s and early 1880s.[28] And it was in 1878, whilst he was on a trip to France that Huxley had helped arrange, that Geddes conducted the first of what would turn out to be a series of studies on the taxonomic differences between plants and animals, and more generally on the nature of biological autonomy.[29] In pursuing these topics Geddes joined a debate that was focused on two sets of connected questions. The first had been prompted by observations made with the aid of increasingly powerful microscopes, which had revealed that some organic entities long assumed to be singular were in fact associations of a number of organisms, one example being the union of fungus and alga that creates lichen. The question that was asked of those relationships was how they should be described, given that they were so close that the organisms could not be separated. Many mid-nineteenth-century men of science, including the Swiss botanist Simon Schwendener, used master/slave analogies, where one organism controls the other for its own selfish ends. By the late 1870s, however, simplistic models based on domination or parasitism were being dismissed as inadequate with increasing regularity. Instead, the word 'symbiosis', coined by the Germany-based experimental botanist Anton de Barry in 1877, had emerged as an alternative term for the close union of a number of organisms.

The second set of questions concerned chlorophyll. Long-thought to be found exclusively in plants, chlorophyll had recently been discovered in simple animals. These observations challenged men of science to explain whether or not this animal chlorophyll was of the same kind found in plants, and if it was, whether it served the same purpose. The answers to these questions were thought to hold potentially profound consequences for the biological sciences, as they touched on deep and fundamental issues regarding where the taxonomic line between plants and animals should be drawn. Consequently, many of the most important contemporary scientific thinkers and writers, including Huxley, had become involved in the ensuing discussion.[30]

It was in 1878, whilst working at a marine biological station at Roscoff, on the French coast, that Geddes, as he later wrote, had 'the good fortune to find material for the solution' of these problems in a type of flatworm, 'the chlorophyll-green Planarian'.[31] He had observed, he explained, how 'in fine weather' the sea-dwelling flatworms would congregate close to the surface of the water, 'apparently to bask in the sun', suggesting that 'their chlorophyll...must have its ordinary vegetable functions'.[32] To test this hypothesis, Geddes first designed an experiment in which he used 'a couple of the round shallow glass dishes used in the laboratory as small aquaria'. 'Into the larger vessel', he explained, 'were put Planarians enough to cover the bottom; it was then gently sunk in the pneumatic trough (a tub of sea water), and the smaller [dish], also full, inverted into it. The apparatus was then placed on a shelf in the sunshine', where the flatworms gradually evolved a gas that he was able to show was oxygen.[33] Encouraged by these results, Geddes then subjected the planarians to a chemical experiment to determine the nature of the colouration residing within their tissues. Treating the flatworms with alcohol, he was able to prove two things. Firstly, having dissolved out the chlorophyll from the planarian, he was able to confirm that the colour spectrum it gave as a reaction to the alcohol was closely similar to that produced by the chlorophyll found in plants. However, after 'repeated treatment with alcohol and ether', which culminated in Geddes boiling the 'residue of the Planarians' with water and adding iodine to what was left, he was able to make an even more precise claim. The 'deep blue colouration' of the remains, 'which disappeared on heating, and reappeared on cooling', indicated 'the presence in quantity', he wrote, 'of ordinary vegetable starch'.[34] Thus, on account of these behavioural and chemical similarities, he concluded that 'these Planarians may not unfairly be called Vegetating Animals, for the one case is the precise reciprocal of the other'.[35]

Over the following two years, which included a period spent at a marine station in Naples during 1879, Geddes continued to work on issues connected with these findings and to think more deeply about the nature of the relationships between closely allied organisms. His most important research in this respect came in 1881 when he turned to the widely disputed function of the starch-containing yellow cells found in radiolarians – a simple marine organism that Ernst Haeckel's work, in particular his extraordinarily detailed copper-plate etchings, had made famous in the early 1860s.[36] (See Figure 3.1.) What made these yellow cells, which Geddes identified as algae, so interesting was that they not only presented the same vegetative signs he had observed in the planarians but also that they could be successfully separated from the radiolarians. In Geddes' view, scientific orthodoxy, which interpreted the cells parasitically, failed to explain 'the mode of life and...the function of such organisms'.[37] The association could not be

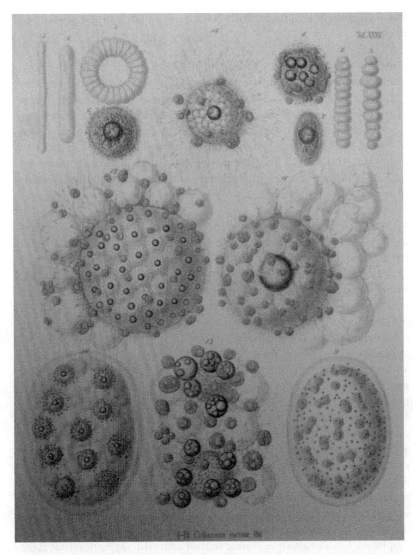

Figure 3.1 Ernst Haeckel's copper-plate etching of Collozoum inerme – a radio-larian that Geddes used in his experiments in Naples in late 1881
Source: Haekel 1862: ii. Plate 35.

parasitic, he argued, because if it were 'the animals so infested would be weakened, whereas their exceptional success in the struggle for existence' was made clear by their outnumbering those radiolarians without algae. Instead, it was more plausible, he suggested, to explain the arrangement as a mutually beneficial one. 'For a vegetable cell', he wrote:

> no more ideal existence can be imagined than that within the body of an animal cell of sufficient active vitality to manure it with carbonic acid and nitrogen waste, yet of sufficient transparency to allow the free entrance of the necessary light. And conversely, for an animal cell there can be no more ideal existence than to contain a vegetable cell, constantly removing its waste products, supplying it with oxygen and starch, and being digestible after death...In short, we have here the economic interactions of the animal and the vegetable world reduced to the simplest and closest conceivable form.[38]

Animal and plant lived together cooperatively, Geddes argued, and he used the term 'reciprocal accommodation' to describe the way that, on his account, they had evolved through adjusting to one another.

In late 1881 Geddes was awarded the Ellis Physiology Prize by the University of Edinburgh for his reciprocal accommodation work, which was subsequently published in *Nature*.[39] Unexpectedly, however, the publication provoked a controversy that, in turn, forced Geddes to reveal the Spencerian credentials of his theory. Immediately after the appearance of Geddes' paper, H. N. Moseley, Linacre chair of human and comparative anatomy at Oxford, pointed out in a letter to *Nature* that Geddes' claim to priority in his prize-winning research appeared to have been undermined by the German zoologist Karl Brandt.[40] Only a few months earlier, Brandt had published a paper in which, based on observations of organisms creating unions of equal benefit, he had argued for a mutually exploitative form of symbiosis – a thesis that some perceived to be the same as Geddes'.[41] Despite the apparent similarities with his own work, Geddes was nevertheless highly critical of Brandt's research, which he regarded as 'an extraordinarily slender foundation for the doctrine of "symbiosis"'.[42] Whilst he complained that Brandt's work was riddled with errors, including factual mistakes and misplaced emphases, Geddes also questioned Brandt's specific suggestion that algae and animal could be separated in a way that permitted both to thrive. Although he had found that the organisms could survive apart, Geddes had also observed that they could not flourish; failing, for example, to reproduce themselves. In his view, one could plausibly argue, as Brandt had, that the relationship had been entered into as a mutually advantageous one. Through time, though, a division of labour between animal and plant had clearly evolved, Geddes asserted, in which both had become dependent on the co-operative behaviour of the other – a Spencerian scenario in all but name. Furthermore, in criticizing Brandt, Geddes also revealed that he had reached his own conclusions about the reciprocally accommodating nature of such relationships when he was studying the planarians in Roscoff in 1878. But for the necessity of acquiring basic experimental evidence to ward off critics, Geddes continued, he 'might have published the doctrine of reciprocal accommodation before going to Naples [in 1879] at all'.[43] In other words, reciprocal accommodation was only to be expected in the organic world, or at least to anyone as steeped in Spencer's writings as Geddes was.

Contributing as it did to important scientific debates, Geddes' reciprocal accommodation research helped him establish a high reputation within the scientific community. After reading 'several of [Geddes'] biological papers with very great interest, and ... [forming] ... a high opinion of [his] abilities', Charles Darwin had come to the conclusion, as he put it in a letter to him in 1882, that Geddes was capable of advancing 'knowledge in several branches of science'.[44] August Weismann too felt able to write, in a letter of recommendation, that 'from the standpoint of general biology I can say that Mr. Geddes ranks among those of the living English [*sic*] investigators who have most deeply thought out the general biological problems which equally concern both the vegetable and the animal kingdom'.[45] However, the career as a laboratory-based experimental biologist that Geddes seemed destined to pursue was over almost as soon as it had begun. In late 1879, during a BAAS-funded research trip to Mexico, he had fallen suddenly and mysteriously ill. 'Aggravated by very severe, and as it afterwards turned out, mistaken medical treatment', Geddes had been 'confined to [his] room for upwards of two months and left ... utterly enfeebled'.[46] Moreover, on his return to the UK in early 1880, he found that any strain on his eyesight resulted in severe headaches, which seriously compromised his ability to use microscopes. As a consequence, Geddes was forced to abandon laboratory work altogether by the mid-1880s and explore other ways of making use of what he had learned from Huxley, Spencer, and his experimental practices. Yet, as we will now see, whilst Geddes' illness helps explain the timing of his move away from the laboratory, it does not explain why the subject he decided to pursue first was the relationship between economics, statistics, and sociology.

Geddes' first ventures into the social sciences

After a brief period as a demonstrator at the University of Aberdeen, Geddes decided to settle back in Edinburgh during the early 1880s and search for a job. However, on account of having abandoned degree courses at both the University of Edinburgh and the School of Mines in the 1870s, he found it difficult, despite the success of his reciprocal accommodation research, to obtain permanent scientific employment. As a consequence, he spent most of the 1880s in a series of temporary jobs, most notably as an assistant to Alexander Dickson, who was professor of botany at the University of Edinburgh. Whilst this relatively modest employment meant Geddes often experienced financial difficulties during the 1880s, it did not mean he stagnated in an intellectual sense. On the contrary, even though his research-based publications in biology rapidly declined after 1882, the year of his final reciprocal accommodation paper, Geddes found a new outlet for his energies: the debate about sociology, which, we saw in Chapter 1, had been ignited by J. K. Ingram's address to Section F of the BAAS in 1878. As Geddes' contributions to those discussions showed, he believed that the reforms Ingram had called for could be realized only if social scientists replicated a number of the methods and practices he had learned to use as a biologist.

Geddes' first attempts to bring together his biological training and the debate about sociology came in the form of two almost identical presentations on the subjects of economic science and statistics, which he gave to the Royal Society of Edinburgh and Section F of the BAAS in the summer of 1881.[47] As he explained in a paper called 'Economics and Statistics, Viewed from the Standpoint of the Preliminary Sciences', which he read in September of that year to Section F at their meeting in York, Geddes' decision to tackle those subjects had been motivated by a specific set of concerns: the 'great reforms demanded in Mr Ingram's presidential address' of 1878. Indeed, by satisfying 'three out of the four' of those reforms – 'that the study of the economic phenomena of society ought to be systematically combined with that of the other aspects of social existence ... that the excessive tendency to abstraction and to unreal simplifications should be checked ... [and] that the *a priori* deductive method should be changed for the historical' – Geddes wanted to contribute to the larger goal that Ingram had outlined: the enlarging of Section F 'so as to comprehend the whole of sociology'.[48]

As he had earlier told the Royal Society of Edinburgh in a three-part paper, entitled 'On the Classification of Statistics and Its Results', which he had read in March, April, and May 1881, Geddes thought a significant part of what had brought political economists to their knees in the late 1870s was a widespread suspicion of statistics.[49] For some, he argued, statistics could be broadly conceived as a science, whilst for others it was simply a method. Consequently, there was no uniform approach to the field, which not only meant there was scepticism of what exactly statistics proved but also that the search for the laws governing society was being hindered. The first step towards achieving Ingram's reforms was therefore a definition of statistics. Citing a recent discussion at the Statistical Society of London, Geddes suggested statistics should be seen as 'simply a quantitative record of the observed facts or relations in any branch of science'.[50] The reason was that, on his account, the purpose of statistics was to record individual facts in such a way that the social investigator could then use them to interpret larger sets of relations.

However, as he argued at both the Royal Society of Edinburgh and the BAAS, Geddes believed there needed to be a fundamentally different approach to social statistics if they were to throw light on society as a whole. What was required, he told his audiences, was a new framework for organizing information about society, one that was 'natural, not artificial;...capable of complete specialisation...and...the widest generalisation;...universal in application...simple of understanding, and convenient in use'. For such a scheme, the social sciences need look no further, Geddes suggested, than his own professional home, biology. After all, he explained to the Royal Society of Edinburgh and Section F, the biologist had to 'specialise until every member [of a group] is known in the greatest detail, and also...generalise these groups into larger and larger ones'.[51] In this sense, if social scientists adopted the methods of biologists then they would be able to build a basic conception of society that could serve the same purpose for social analysis as taxonomy did for study of the natural world. Indeed, 'as the term society assumes', Geddes argued, 'some general truths must be common to societies of *Formica*, *Apis*, *Castor*, and *Homo* alike – to ant-hill, bee-hive, beaver-dam, and city, and this must therefore underlie our classification of social facts'.[52]

Elaborating on how these biological principles might inform social classification, Geddes suggested to the Royal Society of Edinburgh and Section F that information about society could be organized according to what he termed 'sociological axioms'. These axioms were applicable, he explained, to every society throughout history and were of five basic types:

(A) those relating to the limits of time and space occupied by the society; (B) those relating to the matter and energy utilised by the society; (C) those relating to the organisms composing the society; (D) those relating to the application of the utilised matter and energy by the given society; (E) those relating to the results of the preceding conditions upon the organisms.[53]

CLASSIFICATION OF STATISTICS. SOCIETY .... DATE									
A.—TERRITORY. I. QUANTITATIVE.			**TERRITORY. II. QUALITATIVE.**			**TERRITORY. III. DECREASE.**			
Existent at last recorded time.	Increase.		Unused.	Used.		By social agency.		By geologic agency.	
	By social agency.	By geologic agency.		Unspe-cialised.	Specialised.				
B.—PRODUCTION. **I. α. SOURCES OF ENERGY IN TERRITORY.**			**II. DEVELOPMENT OF ULTIMATE PRODUCTS.**			**III. Loss. (PREMATURE DISSIPATION OF ENERGY AND DISINTEGRATION OF MATTER.)**			
Primitive chemical affinity. / Earth's in-ternal heat. / Earth's rota-tion.	Solar radiation.		Energy.	Exploitation, manu-facture, and movement (trade and transport).	Ultimate products.	Agency.	In		
	Kine-tic.	Potential. / Earth's crust. / Organ-isms.	See Table I. α.			Physical, Biological, Social.	Exploitation, manufacture, movement, remedial effort, & c.		
β. SOURCES OF MATTER USED FOR OTHER PROPERTIES.			Matter.						
Mineral.	Vegetable.	Animal.	See Table I. β.						
C.—ORGANISMS. I. QUANTITATIVE			**ORGANISMS. II. QUALITATIVE**			**ORGANISMS. III. DECREASE**			
Existent at last recorded time.	Increase.		Biological.		Social.	Emigration.		Death.	
	Immigration.	Birth.	Structure.	Function.	Mutual relations.				
C.—ORGANISMS. OCCUPATIONS. I. (OPERATIONS ON MATTER AND ENERGY.)			**OCCUPATIONS. II.** (DIRECT SERVICES TO ORGANISMS.)			**OCCUPATIONS. III.**			
Exploitation.	Manufacture.	Movement..	Of non-cere-bral functions.	Of cerebral functions.	Of co-ordination.	Unem-ployed.	Disabled.	Destruc-tive.	Reme-dial.
D.—PARTITION (MEDIATE AND ULTI-MATE) TO CLASS I.			**PARTITION TO CLASS II.**			**PARTITION TO CLASS III.**			
D.—USE BY CLASS I.			**USE BY CLASS II.**			**USE BY CLASS III.**			
E.—RESULT TO CLASS I.			**RESULT TO CLASS II.**			**RESULT TO CLASS III.**			

Figure 3.2   Geddes' table for organizing social statistics
Source: Geddes 1881: 525.

By constructing a table along these lines, Geddes argued, all kinds of data could be brought together and arranged so that it satisfied the conditions he had outlined earlier (see Figure 3.2). Indeed, such a classification made clear how different aspects of society related to each other and how together they constituted a social whole.[54]

Despite frequently referring to sociology at the Royal Society of Edinburgh and the BAAS in 1881, Geddes said very little about what exactly the proposed new science was and why it would be necessary in light of the reforms of social science he had discussed. However, in a paper entitled 'An Analysis of the Principles of Economics', which he read to the Royal Society of Edinburgh over the course of four of its meetings in March, April, June, and July 1884, Geddes gave a first glimpse of how sociology fitted with the scheme he had outlined.[55] On his account, sociology followed on from the reform of economics and statistics because it was a general science of society that dealt with questions specialist sciences could no longer lay claim to. For example, Geddes argued, quantitative investigations, such as the state-run census, uncovered information about society that was related to specific, rather than general, aspects of its existence. 'All that concerns only the objective and bodily side of a man is purely biological', he explained, 'and this may be summed up for a number of men, looked at simply as a herd or mass, without leaving the field of pure biology'. However, sociology went beyond numbers and material considerations. Sociology 'concerns itself', he went on, 'with individualities of a higher order – with aggregates of men *integrated* into wholes for definite functions; as firm, bank, company, regiment, post office, and only considers the individual components in relation to these'.[56] Thus, according to Geddes, whilst biologists, economists, and others contributed to specific boxes and columns on the table he had constructed, the sociologist studied the aspects of those findings that were somehow more than the sum of their parts.

Although Geddes continued to present on strictly biological subjects throughout the late 1880s, it was this vision of the social sciences remodelled in accordance with his biological sensibilities that occupied an increasing amount of his time.[57] What characterized his writings and presentations from the mid-1880s onwards, though, was an effort to integrate the natural and social sciences not just in terms of methodology but in terms of theory and explanation as well. At the heart of this endeavour were two ambitions. The first, as has already been noted, was the inspiration he took from Ingram's call to create a new science of society that would be known as sociology. However, the second was a Spencerian desire to connect the account of evolution he had given in his reciprocal accommodation research with the subject matter he had subsequently come to identify for sociology. As we will now see, with the assistance of the biologist J. Arthur Thomson, whom he had taught at Edinburgh during the mid-1880s, before Thomson went to study under Haeckel in Germany, Geddes set about explaining how he thought the biological and social worlds fitted together.[58]

Encyclopaedias and the Evolution of Sex: the Geddesian biosocial synthesis in theory

Addressing readers of the *Encyclopaedia Britannica* under the heading 'Variation and Selection' in 1888 and, with Thomson, the readers of the article 'Evolution' in *Chambers' Encyclopaedia* the following year, Geddes explained that general scientific opinion was often at odds with many of the basic assumptions underpinning the Darwinian argument for the creative power of natural selection.[59] Variation, the raw material upon which natural selection worked, was 'more and more frequently regarded', Geddes and Thomson wrote, 'as taking place on a few definite lines...which environment can only bend and colour, and natural selection no more than prune'.[60] As a result, it was highly unlikely, Geddes argued, that evolutionary development came about, as Darwinians argued it did, through the competition between individuals for resources. Geddes therefore gave his readers a profoundly un-Darwinian yet distinctly Spencerian account of evolution that was grounded in what he had observed in the flatworms and radiolarians earlier in the decade. Evolutionary development is 'definitely associated', he explained, 'with an increased measure of subordination of individual competition to reproductive or social ends, and of interspecific competition to co-operative adaptation'.[61] Consequently, we are compelled 'to interpret the general scheme of evolution', Geddes and Thomson wrote, 'as primarily a materialised ethical process underlying all appearance of "a gladiator's show"'.[62] 'The ideal of evolution is thus an Eden', Geddes proclaimed, 'and although competition can never be wholly eliminated, and progress must be asymptotic, it is much for our pure natural history to see no longer struggle, but love, as "creation's final law"'.[63]

Whilst these articles highlighted Geddes' debt to Spencer when it came to explaining the processes by which evolution worked, they also demonstrated how Geddes' understanding of the scope of evolution owed much to Spencer too. Geddes was convinced, as he and Thomson wrote in 1889, that there is a 'fundamental unity of' evolution that can be detected in the way that the 'same principles' can be 'traced [from the simplest forms of life] into the highest "superorganic" phenomena of mind and society'.[64] However, as he explained in a variety of pamphlets, talks, and newspaper articles throughout the late nineteenth century, Geddes was also concerned about the trend for using evolution to dress what were obviously political ideas in scientific clothes. For example, as he explained in a pamphlet of 1888, entitled *Co-Operation Versus Socialism*, it was clear that 'through struggle for existence and super-abundant population and the consequent survival of individual advantages' enthusiasts for free-market capitalism were guilty of presenting arguments that were 'too much a simple reading of the facts of nature in terms of Malthus and the current individualistic economics'.[65] Moreover, their socialist opponents, with whom Geddes might have been expected to identify, had proven themselves to be equally incapable, he thought, of 'furnish[ing] us with any theory of society adequately scientific'. Socialists were guilty, he argued, of simply deploying arguments that were orthodox political economy 'turned inside out'.[66]

What underscored these criticisms of the way others tried to relate evolution and society was *The Evolution of Sex* – Geddes' first book, which he co-wrote with Thomson in 1889. Published in the Contemporary Science series, which was edited by the sexologist Havelock Ellis, *The Evolution of Sex* was rooted in Geddes' deeply Spencerian explanation of sex and reproduction, which he had previously explained under those headings in the *Encyclopaedia Britannica*, as well as in a paper to the more elite audience at the Royal Society of Edinburgh.[67] Aiming to invite 'the criticism of the biological student', but intending to address 'primarily…the general reader or beginner', Geddes and Thomson wrote *The Evolution of Sex* in a textbook style with numerous diagrams, illustrations, and bullet-point summaries. Moreover, in so doing, they sought to relate their biological theories of evolution to some of the most controversial social issues of the day.

Sex is a result, they argued, using terms Geddes had first deployed in his encyclopaedia articles, of a 'constant antithesis' in natural processes between two forces: a constructive state that conserves energy, known as 'anabolic', and a diametrically opposed course of disruption that dissipates energy, known as 'katabolic'.[68] In a process of evolutionary differentiation, Geddes and Thomson wrote, 'male reproduction' is 'associated with preponderating katabolism, and the female with relative anabolism', and it was these two forces that explained every physical, emotional, and social manifestation of sex in the organic world.[69]

The apparent activity of the sperm and passivity of the egg in repro-duction, for example, could be seen as an expression of anabolic and katabolic characteristics in a biological process. Similarly, behavioural differences observed in men and women could be seen as the social articulation of these physiological factors.[70] The dynamics of anabol-ism and katabolism were therefore 'the fundamental characteristic of living matter'. However, that conception was 'incomplete...unless it be remembered that about this "organic see-saw" there blows the wind of the environment, swaying it now to one side, now to the other'.[71] Thus, 'to dispute whether males or females are higher', Geddes and Thomson argued, 'is like disputing the relative superiority of animals or plants. Each is higher in its own way, and the two are complementary'.[72]

In the fourth and final part of the book, these theoretical principles were applied to a host of explicitly social questions, including contracep-tion, which was a *cause célèbre* amongst social reformers and freethink-ers of the late nineteenth century, who saw it as a tool of empowerment, particularly for women.[73] As a consequence, many of the book's readers were uncomfortable with the radical turn taken in the book's closing chapters. Indeed, in the opinion of the reviewer for *Nature*, Geddes and Thomson were 'controversialists from the first page of their book to the last', and it was 'very much to be regretted that [they had] included a dis-cussion of certain social and ethical problems absolutely unconnected with the title of their book'.[74] At the heart of their analysis, though, was a belief that the dynamic tensions at the heart of the evolution-ary process could help create, as they put it, 'a new ethic of the sexes; and this not merely, or even mainly, as an intellectual construction, but as a discipline of life'.[75] Repeating the arguments that Geddes had been honing over the course of the previous decade, he and Thomson implored their audience to see that evolution is a form of development characterized by the emergence of other-regarding actions and feelings, including cooperation and love, which should lead all forms of life to negate the destructive elements of their existence. For this reason, Geddes and Thomson claimed that the use of contraceptives should be seen as a natural and progressive evolutionary act. Indeed, by premis-ing their argument on the recognition that childbirth is a gruelling, intense, and dangerous ordeal that carries greater risk the more often women go through it, Geddes and Thomson presented contraception as a responsible way for humans to protect the anabolic force in evolution by rationally controlling the forces that shape them.[76]

Geddes and Thomson did not suggest, though, that it was possible to eliminate biology as a source of human identity. In fact, as they put it in their most famous turn of phrase, they believed that 'what was decided among the prehistoric Protozoa cannot be annulled by Act of Parliament'.[77] Almost by virtue of its distinctly Spencerian character, this notion is now considered to be conservative and its meaning has been the cause of much debate amongst historians, especially since Mona Caird, a leading figure in the late nineteenth-century 'new woman' movement, cited *The Evolution of Sex* in support of her arguments about the future equality of men and women.[78] However, when it is returned to the context of the general evolutionary theory of which it was part, it is a belief that highlights the extent to which our current political categories fail to capture the intellectual commitments of the biosocial programme Geddes had been developing since the late 1870s. As a Spencerian, Geddes not only believed evolution was responsible for creating the world in which he lived, but also that if evolution was allowed to work through to its natural conclusion then it would result, as he and Thomson put it in *The Evolution of Sex*, in an 'earthly paradise of love'.[79] Yet far from being an idealized intellectual construct, Geddes' biosocial programme had come to include a range of practical commitments too. Assisted by the fact that, in 1889, after almost a decade of trying, he had finally obtained a permanent academic post, a chair of botany at the University of Dundee, which was endowed by his wealthy friend James Martin White, Geddes had initiated a series of projects in Edinburgh during the 1880s and 1890s that he believed would contribute to making the evolutionary progress he had described in *The Evolution of Sex* a reality.[80] Turning now to consider those commitments, we will see that in attempting to turn theory into practice, Geddes not only honed in on the city as the focus for the reformed social science he had described throughout the 1880s but also, with significant help from his supporters, established a physical base to promote it.

Edinburgh's slums and the Outlook Tower: the Geddesian biosocial synthesis in practice

Given how deeply immersed Geddes was in Spencer's evolutionary philosophy, it was no surprise he and Thomson had argued in *The Evolution of Sex* that social progress was dependent on the laws of evolution. However, as his and Thomson's discussion of contraception had demonstrated, Geddes' Spencerism did not mean, as historians have long insisted it should, that it was only by passively submitting to evolution that people would bring progress about. On the contrary, Geddes thought it was possible to actively engage with the laws of evolution so they were guided towards goals they might otherwise fail to reach. What highlighted this fact was how he had been making an increasing number of practical interventions throughout the 1880s and 1890s into the social problems that plagued parts of his home city, Edinburgh. As well as demonstrating what kind of action Geddes believed was necessary to catalyse social improvement, those interventions also culminated at the start of the twentieth century with an attempt to realize the ultimate goal of his ongoing efforts to reform the social sciences – sociology.

Geddes had first indicated his interest in contributing to the solution of Edinburgh's social problems during the mid-1880s, when he had attended the meetings of a group who initially called themselves the Environment Society but in 1885 had become the Edinburgh Social Union. The interest that brought philanthropists and social reformers such as Geddes to the Edinburgh Social Union was the dilapidated state of the city's Old Town. Since the eighteenth century, Edinburgh's increasingly prosperous middle and upper classes had gradually relocated from their homes in the Old Town to houses in the city's much-celebrated New Town. As a consequence, the older parts of Edinburgh, in particular the tenement buildings near the Royal Mile, had developed into slums of unquestionable misery. Indeed, so bad were the problems in the Old Town, that Edinburgh City Council, empowered by an 1867 act of parliament, had become a late nineteenth-century leader in slum clearance.[81]

Inspired by the examples of social reformers such as Octavia Hill, who had had considerable success in London since the 1860s with the 'five percent philanthropy' model of social housing, through which investors were promised a modest rate of return, the members of the Edinburgh Social Union had come together to tackle the Old Town's problems in a distinctly different way to the city's council.[82] By raising funds to buy buildings in slum areas and renovate them, the Union wanted to create a socially conscious property management company that leased homes to poor tenants on strict but favourable terms. Underpinning this strategy was a belief that the slums were not, as some people suggested, lost causes but places that could be improved. Whilst it was true, the Edinburgh Social Union argued, that the slums were permeated by vice, crime, unemployment, and disease, those problems were more a reflection of the despair brought on by the surroundings than they were of the character of the people who lived there. Indeed, as projects such as Hill's showed, an improvement in the physical appearance of the slums could often help bring about a corresponding improvement in the lifestyles of their residents.[83]

Despite his approval of their activities, Geddes did not involve himself with the Edinburgh Social Union for long. Instead, in late 1886, he purchased a flat in James' Court, a tenement block on the city's Royal Mile that had once been home to David Hume, and moved in with his wife, Anna, an enthusiast for Octavia Hill's work, whom he met at a meeting of the Edinburgh Social Union and married six months earlier. In the words of the political economist James Mavor, when the Geddeses arrived, James' Court was 'a spot where unredeemed squalor had reigned for at least half a century'.[84] But Geddes and his wife had moved to the Old Town with a plan to change all that. By painting walls, putting up window boxes, and creating garden areas, Geddes, his wife, and others, many of whom were former or current students of Geddes, set about improving the physical appearance of the tenement block the Geddeses called home. However, Geddes' efforts did not stop with James' Court. Throughout the 1880s and 1890s, he not only purchased a number of other properties around Edinburgh but also invested £1,500 in the building of Ramsey Gardens, a block of co-operative flats. In addition to moving with his family into Ramsey Gardens in 1893, Geddes also used part of the building as an extension of University Hall, the UK's first self-governing student residence, which he had established in 1887. On account of these successes, progressively minded people, including the infamous Russian émigré and author of *Mutual Aid*, Prince Peter Kropotkin, travelled to the Scottish capital throughout the late nineteenth century to consult with the Geddeses and see what they had achieved.

Although Geddes' efforts to improve the Old Town owed a debt to the activities of other social reformers, not least the members of the Edinburgh Social Union, they were undertaken as part of the biosocial programme he had been developing since the early 1880s. Indeed, according to his friend Mavor, Geddes' move to the slums came about because he believed his 'biological training was of value in revealing...that, important as surroundings might be, the inherent factor was not less important, and that the degeneration of surroundings was an index of the degradation of the people who inhabited them'.[85] So that he could analyse the Old Town in these terms, Geddes had come to make use of a particular tool: the social survey, which he believed was a counterpart of the methods he had used to pick apart the mode of living of flatworms and radiolarians in the late 1870s and early 1880s. With the aid of his wife and supporters, Geddes travelled throughout Edinburgh to collect data about every conceivable aspect of its existence, from its spatial layout to its cultural traditions, which he argued was a necessary part of understanding how particular places had come to be as they were in the late nineteenth century.

Whilst these social surveys resonated with the methods of the reformed social science he had outlined during the early 1880s, Geddes' interest in them had initially come from another source: the work of the French social surveyor Frédéric Le Play, about whom Geddes had first heard when he was in Paris in 1878.[86] A mining engineer who had once occupied the chair of metallurgy at the École des Mines in Paris, Le Play had changed direction in the 1840s when he turned his attention to the social and cultural aspects of the communities he had previously examined from a mining perspective. At the heart of this move was his belief that people's identities are closely related to two things: the place they live and their occupation, which he argued give rise to specific kinds of social structures and cultural norms. Profoundly conservative and paternalistic in its conception and motivation, this idea was one Le Play expressed in a famous dictum: *'Lieu – Travail – Famille'* ('Place – Work – Family'). For example, in coastal communities of the nineteenth century, he explained, life was shaped by the fact most men worked at sea. With men away from the home, a gendered division of labour and power came about whereby men dominated economic affairs and women domestic matters, which in turn gave rise to a particular set of social attitudes and structures. In this sense, by studying different types of families in great depth, Le Play believed it was possible to understand how people's lives are shaped by factors relating to the geographic and economic identity of the places they live. Moreover, by studying families within modern urban settings, he thought it was possible to show how industrial society was dissolving traditional social bonds and certainties, which he admired and wanted to recover.[87]

What Geddes found appealing about Le Play's ideas was the way they seemed to dovetail with his biological sensibilities when it came to social science. In particular, Geddes was drawn to the way that Le Play's 'Place – Work – Family' dictum, which Geddes often modified to become 'Place – Work – Folk', seemed to be closely related to the 'Environment – Function – Organism' formula he was familiar with from biology. Thus, as Geddes deepened his involvement in the Old Town during the late nineteenth century, he utilized social surveys as the first stage of a framework of action that was inspired by the ideas about social progress he had described in *The Evolution of Sex*. Having surveyed an area and decided what needed to be done to improve it, Geddes and his supporters would then set about showing local residents what could be achieved through actions such as repairing particular buildings or creating garden areas for children to play in. However, having started this process, Geddes would gradually withdraw his assistance. The reason was his belief that a genuine, long-term transformation of the Old Town could not be imposed on the people who lived there but was something that had to be achieved through more organic means. Improvement and social progress would only come about, he argued, if the local residents were the major agents of change.

Through these projects, Geddes was able to hone his biosocial ideas and methods in such a way that, by the end of the 1890s, he was actively promoting them as a model of sociological enquiry and therefore as the culmination of the reforms of social science that he had been working on since the early 1880s. Nowhere was this fact more apparent than at the 'Outlook Tower', an observatory with a camera obscura that was located between James' Court and Ramsey Gardens, which Geddes purchased in 1892. As well as housing Geddes and Company, a publishing firm that he and his supporters had established, and offices that handled his property portfolio, the Outlook Tower served as a museum and events venue that Charles Zueblin, a University of Chicago sociologist writing in the *American Journal of Sociology* in 1899, called the 'world's first sociological laboratory'.[88] Utilizing the panoramic views of Edinburgh that were on offer from the top of the building, Geddes spent much of the late 1890s organizing a series of interlinked exhibitions over the tower's five floors, which he designed as an introduction to the general framework of his sociology (see Figure 3.3). As he explained to the geography branch, Section E, of the BAAS at its meeting in Bristol in 1898, his efforts at the Outlook Tower had 'arisen from the attempt ... to prepare an encyclopaedia, but ... in *rational* order, exhibiting things in their mutual relations'. For this reason, he told the audience at the BAAS,

> the exhibition of the ground-floor centres round a globe with an outline survey of the main concepts of World-geography – e.g. an incipient collection of maps and illustrative landscapes, an outline of the progress of geographical discovery and of map-making, &c. The first floor is devoted to the geography and history of Europe in correspondingly fuller treatment; the second is set apart for an outline geography and history of the English-speaking world ... On the third storey is ... a corresponding survey of Scotland, viewed at once as an historic and social entity and as an element of greater nationality; while the fourth storey ... is a museum of Edinburgh, though again not without comparison with Scottish and other cities.

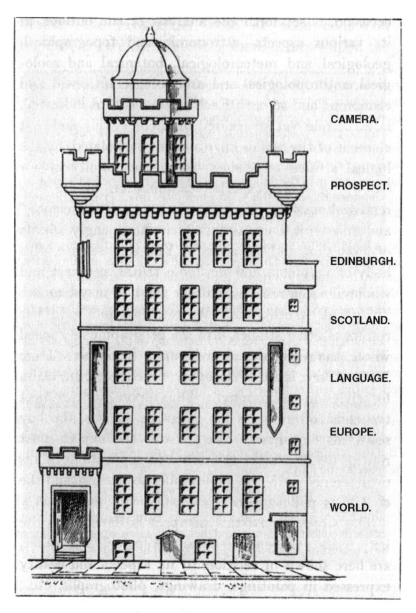

Figure 3.3   A diagram of the Outlook Tower
Source: Geddes 1915: 324.

To maximize the impact of these exhibitions and make the links between them more obvious, Geddes would encourage visitors to begin their tour at the top of the Outlook Tower, where they could see Edinburgh from a new perspective. Then, by 'descending from the roof to the uppermost storey, [the] succession and unity of the physical, organic, and the social conditions [throughout the city would be] better understood'. Yet, as the clear views of buildings such as the neighbouring James' Court and Ramsey Gardens demonstrated, Geddes' intention was not for his visitors to leave with the impression they were being moulded entirely by their environment. On the contrary, he told the BAAS, 'on each level' of the tower 'the view of Nature as determining man [was] complemented by that of man as more or less re-determining Nature' – an approach that was meant to underline Geddes' belief that people could guide the evolutionary forces that shaped them.[89]

However, whilst the ideas Geddes was promoting at the Outlook Tower were his, the fact he was able to do so owed a great deal to the efforts of his supporters. Aside from James Martin White, who had provided him with the security of a job at the University of Dundee, the most important of Geddes' associates in this respect was Victor Branford, whom Geddes had taught at Edinburgh during the mid-1880s. Although Branford had initially harboured ambitions of following in Geddes' footsteps, he had been forced by his family's finances to pursue a more settled line of work.[90] As a consequence, Branford had set himself up in 1893 as the partner in a firm of accountants. For Geddes, Branford's move into the world of business turned out to be crucial because Geddes was on the verge of ruin in the mid-1890s after a decade of continuous property acquisition, which had been financed by a mixture of loans and money his wife, Anna, had inherited from her father. Branford solved these problems for Geddes by creating an organization called the Town and Gown Association, which took over responsibility for Geddes' property dealings in 1896.[91] Thus, freed from many of the financial concerns that had threatened to consume him, Geddes had been able to pursue his development of the Outlook Tower as a home for his sociology.

By the opening decade of the twentieth century, though, Branford had become more than simply a fixer of Geddes' financial problems. Indeed, by acting on his desire to have a career outside of accounting, Branford had become Geddes' chief supporter and a frequent contributor to activities at the Outlook Tower. In the first instance, Branford's efforts in this respect were limited to helping organize events that Geddes had conceived of himself, such as an annual summer school at which members of the public heard lectures from likes of J. Arthur Thomson, the anthropologist A. C. Haddon, and the psychologists Lloyd Morgan and William James. However, in the early years of the 1900s, and in response to Geddes' frustration at being diverted from his projects at the Outlook Tower by teaching in Dundee, Branford had formulated an idea of his own: setting Geddes up as a professional sociologist. In the absence of a sociology department at any British university, Branford realized that the onus was on him and Geddes to engineer a sociology institution of their own. Thus, Geddes and Branford together hatched a plan for a new organization based at the Outlook Tower, which they called the 'Edinburgh School of Sociology'. Although this school was, as Geddes' wife, Anna, put it in a letter of 1902, 'a quite spontaneous movement' that did not extend much further than headed notepaper and occasional meetings at the Outlook Tower, it was something he and Branford hoped would become a fixed part of Edinburgh's intellectual life through which Geddes would find a permanent outlet for his ideas about sociology.[92] The problem, though, and one that we will later see helped catalyse the creation of the Sociological Society, was having the Edinburgh School of Sociology recognized in a way that would enable Geddes to make the transition from being a full-time biologist to a full-time sociologist.

Inspired by the findings of his reciprocal accommodation research and Ingram's call of 1878 for a complete overhaul of the social sciences, Geddes had spent the late nineteenth century developing a comprehensive programme for sociology that was rooted both substantively and methodologically in biology, the field in which he had trained. Moreover, by the opening years of the twentieth century he had come up with a name for that programme: 'civics'. Like the eugenic programme of his contemporary Francis Galton, Geddes had built civics as a system of thought and action that explained not only the state of society as it was in the early 1900s but also how it could be. In this sense, whilst civics and eugenics were connected with biology by different means, with civics being built on explicitly biological foundations and biologists themselves coming to embrace eugenics, both Geddes and Galton offered subtly different but frequently overlapping types of biosocial science. Furthermore, in the process of developing their programme, Geddes and Galton also had arrived in the early twentieth century with supporters who not only wanted to help them promote their ideas but also give them an institutional footing – a fact that would bring them together, we will see in the Chapters 5 and 6, at the Sociological Society. Yet despite the enthusiasm that existed for the kind of biosocial science that Geddes and Galton both offered, not everybody approved of the effort to connect social explanation and biology. As we will now see, another social investigator, L. T. Hobhouse, had been developing a very different framework for sociology, one that was evolutionary but not biological, during the late nineteenth and early twentieth centuries in Oxford, Manchester, and London.

4
L. T. Hobhouse's Evolutionary Philosophy of Reform

'Whether we treat it as biologists, psychologists, or sociologists', wrote Leonard Trelawny Hobhouse (1864–1929) in his book *Democracy and Reaction* in 1904, 'that is to say, whether we compare physical organisms, mental characters, or social institutions, the growth of mind implies always advance of organisation, and advance of organisation depends on the two principles – at first sight opposed – of unity and differentiation'. For example, 'in animals of low organisation', he explained,

> the different parts of the body are so loosely connected as to be in a measure independent of one another. Divide the animal in the proper manner, and you may make of it two, three, or perhaps more separate animals, each quite capable of an independent existence. In some cases the separation often takes place in the natural course, and the observer has difficulty in determining whether he is really dealing with one living being which can easily be divided, or with a number which congregate and act together.[1]

Moreover, Hobhouse went on, the principles of unity and differentiation could be detected in human society, which he held to be the most complex manifestation of evolution, where they 'reappear', he wrote, 'under the familiar names of order and liberty'.[2] As this chapter makes clear, it is only by reconnecting with the philosophical and scientific context in which Hobhouse forged this account of evolutionary development that we can make sense of the reasons he was drawn into the late nineteenth- and early twentieth-century British debate about sociology.

Despite his status as the first, and during his own lifetime only, British professor of sociology, Hobhouse is a thinker and writer about whom historians of science and sociologists know relatively little.

The reason is that Hobhouse, whose 1911 book *Liberalism* is considered a classic of modern political writing, is generally seen as belonging to a political, rather than scientific, tradition of thought.[3] Indeed, for most of the scholars who have written about Hobhouse, it is his association with early twentieth-century 'New Liberalism', the name given to a group of British thinkers, including the radical economist J. A. Hobson, who sought to reconcile the core values of classical liberalism with key aspects of late nineteenth-century socialism, that has proved to be of the most interest.[4] To be sure, the best of these studies, including those by Stefan Collini and Michael Freeden, have always emphasized that Hobhouse was very much a holistic thinker who sought to integrate a range of ideas and practices. When it comes to science, though, Hobhouse has usually been seen as an outsider looking in on debates rather than someone who was an active part of its proceedings.[5]

In recent scholarship, however, a rather different picture of Hobhouse and his work has started to emerge. For example, Gregory Radick, building on the work of Richard Boakes, has returned to the experiments Hobhouse conducted for *Mind in Evolution*, his now seldom read book of 1901, and recovered how Hobhouse actually made significant and important contributions to the field of comparative psychology during the early twentieth century.[6] Moreover, in his survey of the contemporary sociological thought, Steve Fuller has extolled Hobhouse's virtues as a genuinely *social* scientific thinker who endeavoured to protect the autonomy of the human sciences by critically engaging with some of his contemporaries' claims about the importance of biology to social explanation.[7] In this sense, whilst sociologists of the early twenty-first century have yet to experience a resurgence of enthusiasm for Hobhouse as they have for his contemporary, Patrick Geddes, there are signs Hobhouse is once again being taken seriously as thinker in the context of the social and behavioural sciences.

This chapter aims to build on these studies by re-examining the first stage of Hobhouse's career, from his time as a student at Oxford in the late 1880s to the eve of his appointment to the Martin White chair of sociology at the London School of Economics, when he was engaged in the production of a coherent multi-volume project. By exploring each stage of that project, this chapter will chart Hobhouse's emergence as a serious social scientific thinker through his involvement with a number of debates, in particular those regarding how to relate evolutionary theory and philosophy. Firstly, we will see how Hobhouse embarked on the first instalment of his project at Oxford during the late 1880s after deciding, contrary to many of his contemporaries in philosophy, that evolutionary theory did have consequences for his work. We will then see how the second stage of his project was dominated by an effort to develop a theory of evolution that was compatible with the political goals that inspired much of his early writing. As the third section explains, though, an important part of forming that theory was Hobhouse's decision to briefly pause his project during the opening years of the twentieth century whilst he considered whether being an evolutionary thinker meant being a biological thinker too. Having concluded that there was no justification for using biological knowledge to interpret social phenomena, Hobhouse wrote *Morals in Evolution* the final volume of his project, in which, we will see in the closing section, he outlined his vision of an autonomous sociology that was not only free from the constraints of biological thought but was also the means for achieving political and social goals.

What will emerge through this new account of Hobhouse's intellectual development are two points that bear directly on our understanding of his work and the style of sociology he represents. The first, as Radick has recently noted, is the deeply Spencerian character of Hobhouse's project.[8] However, and in keeping with what we have just seen of his contemporary Patrick Geddes, Hobhouse's Spencerism was far removed from what historians have traditionally expected of those who were inspired by the originator of the phrase 'survival of the fittest'. Whilst engaging with the basic framework and principles of the Synthetic Philosophy, Hobhouse's late nineteenth- and early twentieth-century work was shaped by his desire to find an answer to what many commentators saw as the biggest flaw in Herbert Spencer's work: his reluctance to account for the place of human agency in evolution. The second and connected point is that in carving out a space for conscious human action in evolution, Hobhouse managed to square philosophy, science, and politics in a way that is important for our understanding of how those three spheres were related during the period. Indeed, it is only by grasping the way Hobhouse accounted for his distinction between biology and sociology that we can understand how and why he went from teaching philosophy at Oxford in the early 1890s to the debates at the Sociological Society in London fifteen years later.

Liberalism and idealism in Oxford

In 1887, after a distinguished undergraduate career at Corpus Christi College, Oxford, L. T. Hobhouse, then 23 years old, took up a fellowship at neighbouring Merton College with a 'scheme' of research to pursue.[9] Although he might have been expected to fall into the life of gentle contemplation that was typical of a late nineteenth-century Oxford philosophy don, Hobhouse had made rather different plans and, by late 1888, he was studying physiology and biochemistry in the university's Museum Laboratory. In fact, as he explained in a letter in December that year, he was spending 'from five to seven hours a day' cutting up 'frogs, and sheeps' hearts and brains' under the direction of the laboratory's demonstrator J. S. Haldane.[10] By observing how biologists viewed and investigated the organic world, Hobhouse's aim was to build a system of ideas that dealt with the deficiencies of others. Indeed, in putting together his system, Hobhouse later wrote, he was particularly concerned with the way Spencer's 'metaphysical safeguards did not rescue the evolution theory from some of the most unfortunate consequences of a materialistic system'.[11]

Whilst Hobhouse's decision to enter the Museum Laboratory in the late 1880s was somewhat unusual, he was motivated to do so by a set of debates that were then taking place in philosophy. With the emergence of science from what had been known as natural philosophy during the eighteenth and early nineteenth centuries, the field of philosophy had entered a period in which neither its identity nor its authority were as certain as they had been before. For many philosophers, the debates generated by that problem came to be focused on questions about methodology and, in particular, the role of tools such as mathematics and logic in their practices.[12] In late nineteenth-century Britain, the kind of intellectual experimentation those discussions often entailed was expressed most clearly by a sustained, and some might say historically anomalous, wave of enthusiasm for a style of idealist philosophy that was associated with, though by no means enslaved to, the writings of German thinkers, including Immanuel Kant and G. F. W. Hegel.[13] Flourishing from the 1860s onwards at Oxford University, the home of philosophers such as T. H. Green, F. H. Bradley, and Edward Caird, the British idealists sought to challenge the empiricism of John Locke, David Hume, and John Stuart Mill by rejecting the distinction that they, and others like them, had made between mind and matter.

For British philosophers of Hobhouse's generation, the debates stimulated by idealism raised a number of important issues. One of the most significant was the question of whether or not the theory of evolution, which had recently transformed much biological and social thought, should have consequences for philosophy. Whilst there were no hard and fast rules about the sides people took on this matter, the most prominent thinkers advocating the fusion of evolution and philosophy included Herbert Spencer and his followers, as well as a number of sec-ond-generation idealists, such as Bernard Bosanquet and D. G. Ritchie.[14] However, for the philosophers who constituted the first wave of ideal-ism, evolution reeked of the kind of naturalism to which their more spir-itual ideas were opposed. T. H. Green, for example, was bemused by the need of Spencer and others to incorporate vast amounts of biological, psychological, and other scientific observations into their philosophy. Indeed, Green was at a loss, as he once famously put it in response to a question from the political writer and educationist Graham Wallas, to see why his interest in the human mind should also mean that he be interested in the minds of dogs or any other animal.[15]

On the other hand, there were philosophers who objected both to philosophical naturalism and to the elaborate system building of Spencer and the idealists. Amongst the most prominent of these late nineteenth- and early twentieth-century dissenters were Bertrand Russell and G. E. Moore, who thought biological evolution was irrel-evant to philosophy and instead concentrated on developing mathe-matics and logic. Indeed, as he later put it in his book *Our Knowledge of the External World*, Russell believed that 'what biology has rendered probable is that the diverse species arose by adaptation from a less dif-ferentiated ancestry. This fact is in itself exceedingly interesting, but it is not the kind of fact from which philosophical consequences follow'.[16] Promoting a form of philosophy based on the careful examination of propositions and premises, which they called 'analytic', Russell and Moore assembled a range of tools and doctrines, such as the 'naturalis-tic fallacy', which they believed would bring the rigour of the physical sciences to their field.

This maze of philosophical dividing lines was complicated further, though, by the general appeal of idealist doctrines during the late 1870s and early 1880s – a period when, as we saw in Chapter 1, politi-cal and social scientific debate was characterized by a reaction against the abstract and atomistic doctrines of classical political economy.[17] Building on their rejection of the separation between mind and mat-ter, the idealists developed an account of society, generally known as 'social organicism', in which they argued that the interconnections of the social whole make it more than the sum of its parts.[18] As the career of T. H. Green, an active member of the Liberal party and an assistant commissioner on the Taunton Commission of 1864–1867, illustrates, a common result of idealist positions in philosophy was an attempt to reconcile the core of mid-nineteenth-century liberalism, such as its emphasis on individual liberty and rights, with the belief that those values exist in a framework of communal responsibility.[19] Indeed, as the idealist philosopher R. G. Collingwood explained in the early twentieth century, the effect of 'the school of Green' was to send out 'into public life a stream of ex-pupils who carried with them the conviction that philosophy, and in particular the philosophy they had learnt at Oxford, was an important thing, and that their vocation was to put it into practice'.[20]

In formulating the content and substance of his 'scheme', Hobhouse lined up behind both those who believed evolution did make a difference to philosophy and those who were inspired by the political aspirations of idealism. A committed liberal, Hobhouse had mixed with social reformers at university through a range of activities, including his attendance of a discussion group run by the economist Alfred Marshall and the Social Science Club organized by the Fabian socialist Sidney Ball. Moreover, outside of the university, Hobhouse had not only involved himself in the affairs of trades unions around Oxford but also became one of the university's many notable alumni to visit Toynbee Hall, the recently established university settlement in east London, in 1889.[21] It was therefore no surprise that in 1893, whilst he was busily working away on the first instalment of his grand multi-volume project, Hobhouse published *The Labour Movement*, a short book in which he set out his case for state-driven social reform on the grounds of it being a natural progression in the political development of British society.[22]

Citing Green's political philosophy approvingly, Hobhouse explained in *The Labour Movement* his belief that collectivist social movements, such as cooperatives and trades unions, were compatible with traditional political concepts, including individual liberty. However, this reconciliation of what were often seen as opposed political philosophies was only possible, Hobhouse claimed, if purposeful social organization was understood as the most reliable means of achieving individual ends.[23] Collective acts, such as economic intervention to ensure fair wages and prices, were the best way to achieve individual goals, he argued, because 'intelligence is better than blind force, and reaches its end more speedily and surely'.[24] Thus, 'on the ground that intelligence is more effective than brute matter, and that the control of the community is the only possible intelligent agency which can direct the course of economic progress', people were justified in using their collective forms, such as trades unions and, ultimately, the state, to guide society and its growth.[25]

Yet, as he made clear in *The Theory of Knowledge*, the first volume of his project, which was published in 1896, Hobhouse's convergence with the idealists on these points did not mean he thought his contemporaries should 'forget what we have learnt from Mill and Spencer'.[26] As Hobhouse had made clear when he had entered the Museum Laboratory at Oxford, he believed it was essential for philosophers to engage with, rather than diverge from, the sciences. The reason, he argued, was that the sciences presented an account of the world that a great many people found compelling. Thus, in *The Theory of Knowledge*, Hobhouse sought to explain 'the broad, fundamental conditions on which our knowledge and belief in general are founded' – a ground-clearing exercise and foundation stone of a more general series that mirrored Spencer's intentions in *First Principles*.[27] However, if Hobhouse was looking backwards in this respect, he was looking forwards in others. Indeed, as the idealist philosopher J. S. Mackenzie noted in his review for the journal *Mind*, *The Theory of Knowledge* was one of the first English-language studies in the field of epistemology, which had traditionally belonged to logic and metaphysics.[28]

Over the course of more than four hundred pages, which were divided into three parts, Hobhouse explained in *The Theory of Knowledge* how he believed humans to acquire ideas and then go on to validate some of them as knowledge. At the heart of his argument was his claim that there is strong evidence to support the position of empiricists on the existence of a material world that is independent of our senses. However, whilst opposing the idealists on this question, Hobhouse also reached out to them when he suggested that a realist belief in a material world was compatible with idealist claims about the nature of experience. Empiricist philosophers, such as John Locke and David Hume, had made the mistake, Hobhouse argued, of writing about ideas as atomized impressions on the senses. Yet, Hobhouse explained, the human mind can be seen to play an active part in bringing together and making sense of ideas through a variety of its capacities, including what he called 'judgement' and 'inference'. Indeed, he went on, by detecting the relationships between our ideas, we gradually make contact with deeper and more complex levels of reality that require new explanations to unify the whole of which they are a part. We therefore validate an idea as knowledge, Hobhouse claimed, by demonstrating that it fits with both other ideas and our understanding of reality in its entirety. In this sense, he concluded that philosophy had to be a synthetic enterprise in which an engagement with science was a significant component.

Having set out his account of knowledge in its most general sense and given an indication of his overall aims, Hobhouse was ready in 1896 to move on to the second stage of the project he had formulated almost ten years earlier. To meet the challenge of producing volume two of his scheme, though, he felt he needed a change of scenery. However, in searching for a route out of Oxford, Hobhouse did not plan to change the division of intellectual labour that had been characterized by *The Labour Movement* and *The Theory of Knowledge*. Thus, whilst he continued to explore and articulate his political ideas during the late 1890s and the opening years of the twentieth century, he simultaneously set about the more focused task of researching the second instalment of his planned series. As we will now see, the outcome of that effort was an increasing convergence of those two aspects of his work.

The *Manchester Guardian* and *Mind in Evolution*

'As you know', Hobhouse wrote to fellow Oxford don Arthur Sidgwick in late 1896, 'I am thinking of leaving Oxford'.[29] After 13 years at the university, Hobhouse 'longed', as one of his students put it, 'to use his powers in the world of action'.[30] However, as Hobhouse also told Sidgwick, he was clear that, whatever he chose to do next, nothing was to 'interfere with any [of his] own studies'. 'The sort of thing I could do, & should like to do', Hobhouse explained, 'would be to contribute the occasional articles of a somewhat solid & useful character which the [*Manchester*] *Guardian* often prints. Any political, social or economic subject interests me & I would be ready to try my hand at any such topic that wanted treatment'.[31] Recommending him as 'a strong Liberal & progressive of the best type', Sidgwick forwarded Hobhouse's letter to C. P. Scott, the editor of the *Guardian*, and, after a brief trial, Scott granted Hobhouse his wish by offering him a job that required him to attend the office only in the afternoons.[32]

That Hobhouse wanted to write for the *Guardian* was unsurprising. Founded in 1821 at the geographic heart of British manufacturing and industry, the newspaper had made its name as a Whig publication representing the interests of both trade and religious and political dissenters.[33] However, under the editorship of Scott, whose tenure began in 1872, the *Guardian* had changed direction by engaging with the kinds of questions that had captivated the minds of British social reformers since the decline of classical political economy and other narrowly individualistic doctrines during the late 1870s. Moreover, after the split of the parliamentary Liberals into Liberal and Liberal-Unionist parties over the issue of Irish Home Rule in the mid-1880s, the *Guardian* had come to assume even greater importance in radical and liberal political culture. In fact, from the mid-1890s onwards, the newspaper 'was Liberalism', Hobhouse later wrote, because 'it was to the paper rather than to any personal leader that the thoughtful Liberal looked for stability of purpose'.[34]

Despite the heavy workload that came with organizing his days around the research for his project in the mornings and journalism in the afternoons, Hobhouse was renowned for his prodigious writing output, which was evidenced by the 322 articles he produced for the *Guardian* between 1901 and 1902 alone.[35] Although his articles were of a variety of different types, he was primarily a political journalist for the newspaper and his immense productivity was partly a consequence of the fact that the Conservatives' parliamentary advantage over the Liberals gave him no shortage of issues to sink his teeth into. Yet, whilst domestic matters, such as trade, education, and industry, were all the subjects of Hobhouse's writing, it was foreign affairs and the Boer War (1899–1902) in particular that captured both his and the *Guardian*'s imagination during the closing years of the nineteenth and opening of the twentieth centuries.

Described by Hobhouse as the 'test issue of this generation', the Boer War further split the already fractured British liberal and leftwing communities, which struggled not only with the general justification for the war but also with the hugely controversial methods that the British government had pursued, including scorched earth tactics and concentration camps.[36] With its strident opposition to the war and vocal objections to the jingoism of its supporters, the *Guardian* was positioned on 'the unpopular side' of the debate and for Hobhouse, whose sister, Emily, was a member of the South African Conciliation Committee, and from whose pen the newspaper's anti-war leaders often came, it was impossible to forgive anyone who had supported it.[37] As he explained in an article entitled 'The Ethical Basis of Collectivism', which was published in the *International Journal of Ethics* in 1899, the kind of collective organization he and other social reformers advocated was a profoundly ethical endeavour with an international dimension. Collectivism involved 'the substitution of the principle of peace on earth and good-will towards men for the principle of rivalry and war'. 'It thus aims', he wrote,

> at extending a process which has always been going on so far as moral progress has been a reality, that of qualifying the methods of competition of the social spirit realized in the more intimate personal relations. And in so doing it reveals itself as a phase in a still wider evolutionary process, whereby in the higher races purposive and intelligent organization of life, inclusive of the whole species is gradually substituted for the war of all against all by which the survival of types is determined at a lower stage.[38]

Whilst this belief inspired much of Hobhouse's journalism and political writing, it had its roots, as the reference to evolution suggests, in the ideas he had been exploring during the construction of *Mind in Evolution*: the second stage of his philosophical project, which was published in 1901. At the heart of that book was a 'general conception of mental evolution', which, he admitted in the preface, he had 'formed some fourteen or fifteen years' earlier.[39] However, to give empirical rigour to that theory he had designed a series of experiments, which he had the opportunity to carry out in late 1900 during his mornings off from the *Guardian*. Hobhouse's aim, he explained, was 'to test [his] hypothesis so far as animal intelligence and the generic distinction between animal and human intelligence are concerned'.[40] Yet far from attempting to underscore the boundary between human and animal minds, which was maintained by many thinkers, Hobhouse wanted to blur the lines between the two. More specifically, he wanted to show how the ability to shape and guide evolution, which he had written about in both *The Labour Movement* and *The Theory of Knowledge*, was something that emerged gradually in the natural world.

The experiments Hobhouse had designed to achieve this end owed a great deal to the recent work of the psychologists Edward Thorndike and C. Lloyd Morgan.[41] In essence, Thorndike and Morgan had used puzzle-box-style experiments to test the learning and reasoning capacities of animals. In so doing, they had argued there was little evidence to suggest that animals possess what we would call intelligent faculties. Building on their work, Hobhouse had designed a similar set of experiments that he carried out first 'with [his] own dog and cat' and then with 'several animals in [the] great collection at the Belle Vue Gardens, Manchester, including monkeys of several species, a young female elephant, and an otter'.[42] Over the course of several months, Hobhouse had set his animal subjects 'the task of obtaining food by some method presumably strange to it. For example', he wrote,

> food was put into a box, which was then shut, and left for the animal to open, or it was placed out of reach, yet so that it could be obtained by pulling a string, or pushing a door. The animal was first allowed time to discover the method of obtaining it for itself. If after a little while it showed no sign of hitting on the right method, it was shown, and allowed to get the food. Fresh food was then placed as before, and a new trial begun.[43]

Hobhouse's intention was for these experiments to 'measure the influence of perceptual acquisition (learning by perception of results) as

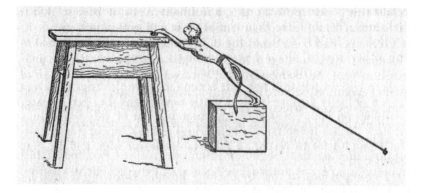

Figure 4.1   Diagram of one of Hobhouse's experiments with a Rhesus monkey called Jimmy

Note: In this case, Hobhouse placed a 'piece of potato on a table at such a distance from the point to which his cord was fastened that he could just reach it if [, by using the box as a stool,] his body and the cord made as nearly as possible a straight line.' Jimmy's 'success in this experiment indicates a power of fairly precise adjustment of one thing to another; and it is one in which neither accident nor imitation can be supposed to have played much part.'

Source: Hobhouse 1901: 248–51.

distinct from motor acquisition (learning an act by doing it)'.[44] He therefore tried to ensure that the animals could not stumble upon the solutions to problems accidentally – a fault he had specifically identified with Thorndike's methods.[45] Thus, in one experiment, which was carried out with a monkey at Belle Vue Gardens, Hobhouse positioned pieces of banana around the animal's cage in such a way that they could only be reached with the aid of a long stick. However, to eliminate the likelihood of the monkey obtaining the banana by chance, Hobhouse made it necessary for the longer stick to be acquired with the aid of a shorter one.[46] The ability of many of the animals to meet challenges such as these demonstrated they were capable of insight learning; that is, of something more than instinctual thought. Indeed, the variation in the animals' abilities to solve the problems was proof, he argued, that the ability to organize facts and experiences became more pronounced at each new stage of development.[47]

Hobhouse called this view of evolution 'Orthogenic' and he described it as 'the unfolding of all that there is of latent possibility in Mind, the awakening of its powers, the development of its scope'.[48] Moreover, he argued that this recognition of the gradual emergence of intelligence should have significant implications for the general perception of evolution as a whole. Whilst 'organisation is at first physical or biological', Hobhouse explained, 'the development of intelligence consists in widening the scope of this function, as well as in perfecting its execution'. For this reason, he went on, although 'life is indeed in a sense organised without intelligence', it is only so 'in a rudimentary way' because 'it is so built as to behave itself appropriately within a certain groove'. When trapped in that groove, Hobhouse argued, organisms are subject to the fierce and largely mechanical laws, such as the struggle for existence, which were largely associated with biology. In that state, a race or species can 'never ... [be] master of its surroundings', he wrote, because it is generally subject to laws of which it is not aware. It was in this sense, Hobhouse argued, that 'the growth of mind works a gradual revolution' because, as organisms learn about the forces shaping them, they and, indeed, we acquire the ability to 'reorganis[e] life on the basis of [that] knowledge'.[49]

According to Hobhouse, one of the most important products of these changes was the advent of purpose. By consciously formulating goals and strategies to achieve them, rather than being slaves to impulsive instincts, organisms entered a new phase of life. As Spencer and others had documented, this new stage was based 'not on internecine rivalry', Hobhouse observed, 'but on mutual interdependence'.[50] Thus, 'the conclusion to which we have been led', Hobhouse went on,

is that among the manifold conflicting movements of evolution, there is one tendency of which the significance is not obscure. In orthogenic evolution we find a constant development of Mind in scope, and accordingly in power. Slow at first, the development gathers speed with growth, and finally settles in to the steady movement of a germ unfolding under the direction of an intelligent knowledge of its powers and of its life-conditions. The goal of the movement, as far as we can foresee at present, is the mastery by the human mind of the conditions, internal as well as external, of its life and growth. The primitive intelligence is useful to the organism as a more elastic method of adjusting itself to its environment. As the mental powers develop, the tables are turned, and the mind adjusts its environment to its own needs...But the last enemy that man shall overcome is himself. The internal conditions of life, the physiological basis of mental activity, the sociological laws that operate for the most part unconsciously, are parts of the 'environment' which the self-conscious intelligence has to master, and it is on this mastery that the *regnum hominis* will rest.[51]

Coming as it did in the closing pages of *Mind in Evolution*, Hobhouse's reference to 'sociological laws' was important because it hinted at what the next step in his Spencerian project was going to be. Having covered philosophical first principles in *The Theory of Knowledge*, and psychology in *Mind in Evolution*, Hobhouse was shifting his attention at the end of 1901 to the challenge of outlining the social manifestations of orthogenic evolution. However, to do so, he had to first make clear the ways in which biology was to feature in his method of social explanation. Whilst he had, to a certain extent, already outlined some of his ideas about this issue in *Mind in Evolution*, the growing popularity of biosocial science, such as the programme of eugenics promoted by Francis Galton, at the start of the twentieth century made it crucial Hobhouse set out his position on the subject. Thus, before producing what he thought was a genuinely *social* science, conforming to the principles he had described in his work so far, he took time out to critique what he saw as the falsehoods of social science that was enthralled to biology.

Social science and the biological challenge to social reform

In 1901, after almost four years in Manchester juggling the twin demands of research and journalism, Hobhouse believed his time with the *Guardian* was coming to an end. As he put it in a letter to Scott, Hobhouse had become 'pretty much convinced that [he could not] *permanently*' carry on with the arrangement and that 'philosophy [had] the first claim on [him]'.[52] Hobhouse therefore ended his full-time association with the *Guardian* in 1902 and moved to London, where he continued with his research and took on a number of different jobs.[53] Whilst contributing to the political press and, on occasions, lecturing to university audiences in Birmingham and London, Hobhouse also took up the post of secretary of the Free Trade Union, an organization that was opposed to the Chancellor of the Exchequer Joseph Chamberlain's plan to raise revenue for the cash-strapped government by reforming trade tariffs. Throughout this period, however, Hobhouse's major concern was to deal with a specific set of issues that related directly to the next volume of his project in which he planned to move his evolutionary analysis from individual minds to society as a whole.

Although the timing of his focus on social science, and sociology in particular, was driven by the momentum of his research, it came at what was, practically speaking, an opportune moment. In June 1903, the year of Spencer's death, Hobhouse was amongst the group of over 50 people who attended the conference held at the Royal Statistical Society in London to 'consider whether the opportunity was suitable for the formation of a Society to promote sociological studies'.[54] When the Sociological Society was officially formed in November 1903, a set of events that will be examined in the two chapters that follow, Hobhouse was elected to the Society's council and joined the editorial committee that helped prepare the Society's accounts of its meetings, which were known as the *Sociological Papers*. Moreover, he was also one of a number of thinkers, including Patrick Geddes and the anthropologists Edward Westermarck and A. C. Haddon, who contributed to the Sociological Society's efforts to promote their activities by giving a course of lectures on their work at the University of London. In so doing, Hobhouse was at the geographical and intellectual heart of the British sociology movement at precisely the moment when he was thinking about what was the correct way, scientifically speaking, to talk about society.

As he had suggested in the closing chapters of *Mind in Evolution*, Hobhouse was convinced that the proper study of psychological and social phenomena required a nuanced understanding of how those fields were demarcated from biology. This fact seemed especially important to him because of what he saw as a dangerous tendency amongst some of his contemporaries to reduce complex political and social problems to seemingly simple biological ideas. What Hobhouse had in mind were issues such as degeneration and national efficiency, which were not only rooted in biological metaphors and analogies but had also come to dominate British political debate during the first five years of the twentieth century.[55] With the weight of biosocial science, such as eugenics, increasingly being brought to bear on political debate, he believed that there was a need, as he put it in a review for the *Guardian* of the American sociologist Lester Ward's book, *Pure Sociology*, to 'liberate the science of society from the grip of the biologists, or rather to put the two sciences into their proper relation'.[56]

Hobhouse's motivation for scrutinizing biological claims about society was ultimately political. As he put it in his 1904 book *Democracy and Reaction*, 'the doctrine that human progress depends upon the forces which condition all biological evolution' was frequently used as a defence of the kind of conservative political doctrines that he opposed. In fact, 'just as the doctrine of Malthus was the main theoretical obstacle to all schemes of social progress through the first two-thirds of the [nineteenth] century', Hobhouse wrote, 'so the doctrine derived in part from Malthus by Darwin has provided a philosophy for the reaction of the last third'.[57] For this reason, Hobhouse believed that the social sciences were a crucial part of the social reformer's toolbox. Indeed, as he had indicated in his work up until that point, Hobhouse was of the opinion that the social sciences would enable humans to understand the forces that shaped them and, as a consequence, make his reformist vision a reality.

To overcome biological arguments against positive social reform, Hobhouse believed it was necessary to critically engage with, rather than avoid, the claims being made on behalf of biologists. Whilst it was true, he conceded, that 'every higher and more special science is in part dependent on those which are lower' and that, in this sense, 'every science that deals with man must take account of the results ascertained by biology as the general science of life', he did not think biology had the final say on social matters. After all, 'since it deals with organised matter', Hobhouse wrote, '[biology] must learn from physics and chemistry what they have to teach about the behaviour of matter in general and of the specific substances found in organised bodies in particular'. However, he went on,

> when the biologist comes to deal with the actual behaviour of organic matter in the living organism, he is by no means disposed to let the physicist or chemist dogmatise as to what he must find. On the contrary he knows what he himself finds by his own methods, and this is often enough the very opposite of what pure physics would lead him to expect. The higher science, ... though dependent on the lower, ought by no means to merge its identity therein.[58]

For Hobhouse, the line dividing biology and the social sciences was to be drawn at the point where mechanical processes, including the struggle for existence, were transformed by conscious ones, such as meaning and purpose. Through systems such as moral codes and traditions, as well as the generally recognized increase in mutual aid, society was defined, he argued, by forms of organization that went far beyond those studied by biologists. Such purposeful development, which could be explained as a collective effort on the part of humanity to overcome the material forces that constrain it, was, he argued in a chapter of *Democracy and Reaction* entitled 'Evolution and Sociology', something biology was unable to explain. 'There is no inherent "upward" tendency in evolution', he wrote, 'so far as it is dependent on the struggle for existence. Old types may be maintained, or new types may arise, but there is nothing to determine whether the new will or will not be capable of a life fuller and better ... than the old'. In lacking this qualitative or meaningful dimension, he argued, biological laws, which were being touted by many as the solution to social problems, 'have no essential tendency to make for human progress'.[59]

An excellent example of someone who conflated biological and social explanation in just this kind of way was, Hobhouse observed, Karl Pearson – professor of statistics at University College London, and, as we saw in Chapter 2, a prominent supporter of Francis Galton. In a pamphlet of 1901, entitled *National Life from the Standpoint of Science*, Pearson had put forward an argument that was becoming increasingly popular in the national efficiency movement, which used Darwinism to interpret social progress as resulting from a conflict between nations, rather than individuals.[60] 'History shows me one way', Pearson wrote, 'in which a high state of civilization has been produced, namely, the struggle of race with race, and the survival of the physically and mentally fitter race'.[61] In Hobhouse's view, there seemed to be a fault in the way Pearson came to this conclusion. The races Pearson saw as incapable of 'producing a civilisation in the least comparable with [the] Aryan' lived, Hobhouse observed, 'in small communities in which the state of war may be said to be normal' – a point that even Pearson recognised.[62] If it was the case that struggle caused social development, Hobhouse asked, then 'why does not the savage, whose struggle is the keener, progress more rapidly than the civilised race, where the struggle is mitigated?'[63]

Of course, it was one thing for Hobhouse to criticize the biosocial science of others and quite another for him to provide a positive sense of what social scientific explanation should be like. Whilst he was not yet in a position to publish the volume of his project that would do this, he did use *Democracy and Reaction* to provide a hint of what his readers could expect from the follow up to *Mind in Evolution*. Indeed, by building on his arguments about how 'a just conception of evolution ... does not support the view that the struggle for existence is the condition of progress', Hobhouse outlined the 'ethical conception of progress' that he believed should guide social science and was rooted in his understanding of the 'evolution of mind, that is to say, in the unfolding of an order of ideas by which life is stimulated and guided'.[64] 'There arises by degrees', he wrote, 'the ideal of collective humanity, self-determining in its progress, as the supreme object of human activity, and the final standard by which the laws of conduct should be judged'.[65] In this sense, he argued, sociology, as the general science of society, was more than a descriptive science; it was a guide for social action. Indeed, whereas a science of society based almost entirely on biological ideas was 'profoundly reactionary', Hobhouse wrote, a 'truer view of evolution ... exhibits the attempt to remodel society by a reasoned conception of social justice as precisely the movement required at the present stage of the growth of mind'.[66]

With this basic account of human society and the purpose of sociology in place, Hobhouse set about writing *Morals in Evolution*, the final instalment of his scheme, during 1905. By shifting the focus of Hobhouse's analysis to social development, *Morals in Evolution* was meant to provide a synthesis of the theory of evolution he had set out in *Mind in Evolution*, the criticisms of biosocial science he had been articulating during the opening years of the twentieth century, and the political goals that had inspired him since the late 1880s. As we will now see, the outcome of that effort highlighted a number of important points about the nature of Hobhouse's development over the course of the previous two decades. Indeed, whilst highlighting how far he had diverged from his starting point in philosophy, *Morals in Evolution* also certified Hobhouse's credentials as a serious social scientific thinker in such a way that in 1907, when he found himself without a job, he was immediately considered a candidate for the first British chair of sociology.

Morals and sociology

'Although Mr. Hobhouse's mastery is undoubtedly superior to Herbert Spencer's', wrote the philosopher James Seth in the *International Journal of Ethics* in 1908, 'we are too often reminded [in *Morals in Evolution*] of Spencer's miscellaneous assortment of facts and too rapid generalization from them'.[67] Whilst Seth's comparison of Hobhouse with Spencer captured what Seth described as 'the broad sweep of [Hobhouse's] range of acquaintance with the facts and his grasp of their essential significance and comparative interest', it was apt in other ways too. Having dealt with philosophical first principles in *The Theory of Knowledge*, psychology in *Mind in Evolution*, and biology in his journalism and *Democracy and Reaction*, Hobhouse brought his two decade-old intellectual undertaking to an end in 1906 by dealing with sociology and ethics – the subjects of the final two parts of the Synthetic Philosophy. However, in so doing, Hobhouse did not simply cycle through the points that had brought Spencer's project to an end in the early 1890s. On the contrary, Hobhouse fused sociology and ethics in a way that provided a robust response to those who maintained that evolutionary theories led to passivity towards social affairs. In this sense, by extending his account of intelligence and rational control from individual minds to society as a whole, *Morals in Evolution* was Hobhouse's solution to the problems with

Spencer's evolutionary philosophy, which had originally helped inspire his work during the late 1880s.

As his choice of title indicated, Hobhouse's aim in *Morals in Evolution* was to relate ethics and social development; an issue he had first discussed in 1904 in his lectures at the University of London, which, as he told the philosopher Lady Victoria Welby, had been 'a study of historical facts out of which [he]...hoped ultimately to extract...[a] philosophic theory of evolution'.[68] Building on those lectures, Hobhouse had divided *Morals in Evolution* into two volumes. In the first, he conducted a general survey of ethics throughout history by 'distinguish[ing] and classify[ing] different forms of ethical ideas' in a way that was 'comparable', he wrote, 'to the physical morphology of animals and plants'.[69] Examining areas such as the law, marriage, property, and class relations, he traced how ethics had changed and developed from the earliest 'savage' or 'primitive' peoples through to 'civilized' societies, which in Hobhouse's view began in the era of the Babylonian and Egyptian empires and ran right through to his own. The second volume was a history of ideas, mainly of religion and philosophy, which Hobhouse saw as providing the justifications for the different types of social system he had explored in volume one. As he had indicated in his letter to Lady Welby, the point of this process was to build an account of human history in which modern society was seen to have grown out of earlier cultures. More specifically, though, what Hobhouse wanted to argue was that the personal freedoms of modern societies, in particular those of early twentieth-century Britain, were things that had rationally evolved out of lower forms of social organization in which individuals were subsumed completely within groups.

The reason Hobhouse gave for focusing on morals was his belief that 'the formation of [ethical] rules, resting as it does on the power of framing and applying general conceptions, is the prime differentia of human morality from animal behaviour'. Indeed, what made ethical systems important for Hobhouse was not just that they existed but that they are 'handed on from generation to generation'. In this sense, he argued, social traditions are 'the dominating factor in the regulation of human conduct' and therefore a necessary subject of study for anyone interested in understanding the relationship between humans and the societies in which they live.[70] However, far from posing a problem for the account of orthogenic evolution that framed his work, ethical traditions were presented by Hobhouse as its highest and clearest manifestation. The lesson of history was, he wrote, that whilst morals had begun 'in the sphere of the unconscious', they, like everything else, had increasingly become the subjects of human intellectual inquiry and rational control. Whilst early societies took their ethics from previous generations without questioning them, 'as human intelligence expands', Hobhouse argued, the 'underlying conditions of ethical movement are no longer left to work out their effects slowly and indirectly'. In fact, moral and social progress come about, he claimed, because at each new stage of development 'the requirements of social welfare are deliberately taken into account in dealing with new questions, and even established customs and traditions are criticized in the light of experience'.[71]

At the heart of this theory, which was clearly meant to justify the political arguments he had put forward in works such as *The Labour Movement*, was a distinction Hobhouse made between sociological and ethical development. Sociology, he argued, is the study of the forces that not only shape society but also remain unknown to the humans being shaped by them. In a sociological sense, Hobhouse explained, social growth 'may produce a set of institutions of a certain value which no brain created, no human being planned, and which even those who enjoy them do not sufficiently appreciate to maintain them against attack'. However, rather than being unconscious, ethical development came about, he went on, through 'the growth of character or of a reasoned conception of the good'. This distinction was important, he argued, because, 'from the ethical point of view[,] institutions depending on a certain degree of ethical advance are of much more value than precisely similar institutions reached by another road'. On Hobhouse's account, 'the difference' between ethical and sociological institutions was therefore 'likely to emerge in their subsequent history' because, just 'as the non-ethical changes of society affect the standard of conduct, so ethical ideas may in their turn re-act upon social organisation'.[72] In this sense, whilst 'ethical evolution... is not the same thing as Social evolution', 'it is intimately connected with it', Hobhouse argued, since history showed that humans are capable of understanding and then overcoming the forces that had previously constrained them. Thus, he concluded, the history of social progress was the history of humans remodelling the world according to their own rationally conceived ideals.[73]

However, having first conceived of this theory in response to the kinds of criticisms that were often levelled at Spencer, Hobhouse was clear his readers should not take him to be implying that social progress is somehow inevitable. By emphasizing the role of purposive human action in social evolution, Hobhouse argued that although change was something that occurred automatically, progress, in the normative sense of the term, is something that can only come about through the deliberate effort on the part of humans to achieve specific ends. It was for this reason that despite their separation on the one hand, Hobhouse saw the ethical and sociological spheres as slowly moving closer together on the other. Indeed, if ethical progress was about humans changing society based on their understanding of how they were being shaped by it, then sociology was a necessary part of furnishing people with the knowledge that was required to make ethical progress happen. This fact was demonstrated, he argued in the second volume of *Morals in Evolution*, by the way that sociological and the ethical change were gradually 'fuse[d] into one' in the highest forms of social organization. For example, 'the State rests on a measure of Right in the relations of men', Hobhouse explained, choosing an example drawn straight from his political philosophy,

> and is so constituted as to be modifiable by the deliberate act of the community. In the method in which changes are effected...we find a definite evolution from the unconscious and unnoted changes of custom, through the deliberate changes introduced on occasion by the fiat of authority[,] to the organic legislation of the modern world in which at its best there is an effort to determine social progress in accordance with a rational ideal.[74]

It was therefore 'on the possibility of controlling social forces by the aid of social science as perfectly as natural forces are controlled at present by the aid of physical science', Hobhouse argued, 'that the permanent progress of humanity must depend'. Indeed, it was just this point that made sociology, 'a science in its infancy', so important.[75]

Whilst this normative view of social science put him firmly in the same camp as many of his contemporaries, including Patrick Geddes and Francis Galton, it also highlighted how far away from the prevailing trends in philosophy Hobhouse had been led during his ascent from first principles to sociology and ethics. Reviewing *Morals in Evolution* for the *Independent Review*, Bertrand Russell accused Hobhouse of committing an error that, along with G. E. Moore, Russell was making a central plank of analytic philosophy: the naturalistic fallacy. 'The study of past moral systems is useful', Russell wrote,

> as showing that society can survive under institutions which to us seem monstrous, and as illustrating the part played by customs and irrational prejudice in almost all beliefs. In this way, it instils

wholesome doubts and promotes a careful examination of our views, and thus may suggest grounds *against* many cherished ethical dogmas; but it is quite incapable of giving grounds *for* any opinion as to what is desirable. Such an opinion can only validly come from our own perception of what is good, not from the distilled essence of the views of previous ages.[76]

In this sense, 'one cannot doubt', Russell argued, 'that [Hobhouse] regards his anthropological data as merely means to an end, namely to his conclusions as to ethics and politics; and', as a consequence, *Morals in Evolution* was 'rendered unsatisfactory by the very insufficient connection between [Hobhouse's] data and his conclusions'.[77]

Responding to these criticisms, Hobhouse sought, not unfairly, to deny that he was an evolutionary naturalist of the kind Russell had accused him of being. Philosophers such as Russell were set against sociology as a whole, Hobhouse argued, which led them to take the line that 'the sociological method involves certain fallacies, and if one avoids them he is not a sociologist'. However, whilst accepting that sociology raised issues about confusing 'questions of fact with questions of value', Hobhouse neither believed he was guilty of doing so nor conceded the two areas should, or indeed could, be separated in the way Russell suggested.[78] As sociologists, Hobhouse argued, 'we are investigating not molecules, nor limpets, but the whole gamut of human interests...and we, the investigators, are naturally and inevitably charged with prepossessions and interests similar to those that we dissect'.[79] It was therefore obvious that the sociologist had to be aware of the kinds of issues that motivated critics such as Russell. For Hobhouse, though, 'the remedy seem[ed] to lie not in the attempt to carry further the divorce of Sociology from Ethics, but rather [in] bring[ing] the two' closer together. 'We shall best avoid confusing the "is" and the "ought"', he argued,

> if we first distinguish and then compare them. If we once have a clear standard of value in our minds we shall at least know what we mean when we apply terms of praise and blame. We shall not, for example, speak of one type as "higher" than another merely because it is later. Still less shall we go on to argue that the higher type always takes the place of the lower, and that therefore all is working for the best in this best of all possible evolutionary schemes.[80]

Yet, despite holding his own against Russell, Hobhouse made two interrelated points clear in his response to the charges Russell had levelled against *Morals in Evolution*. Having taken his cue from evolution, Spencer, and the political philosophy of the idealists, Hobhouse was neither part of philosophy in the way he had been 20 years earlier nor had any great desire to be so in the future. Hobhouse did think objects, organisms, institutions, and ideas were deeply connected to their history and he thought it necessary to take evolution into account when discussing them. He had little sympathy with the ideas underpinning the analytic philosophy project and, for this reason, he and Russell were talking across one another.[81] Indeed, after undertaking two decades of research for the scheme he had first formulated in Oxford, Hobhouse, as his reply to Russell showed, had come to identify himself with sociology in the early twentieth century and not philosophy.

When Hobhouse left the *Guardian* for London in 1902, he had done so with the expressed intention of finishing his project. However, unlike some people who had formulated similarly ambitious plans, such as Francis Galton, whom he had met at meetings of the Sociological Society, Hobhouse was not a man of independent wealth. On the contrary, Hobhouse, who had a wife and three children by 1906, had worked throughout his time in London. In addition to the lectures and post as the secretary of the Free Trade Union, which have already been noted, he continued to contribute to the *Guardian* and other newspapers and periodicals to supplement his income. To that end, and with the publication of *Morals in Evolution* in sight, Hobhouse had taken a full-time job in late 1905 as the political editor at the *Tribune*, a new daily newspaper that had been established by its proprietor Franklin Thomasson as a voice for liberalism. Yet, after a series of editorial disputes with Thomasson, a liberal of the classical rather than new variety, Hobhouse left the *Tribune*.[82] In January 1907 Hobhouse was therefore not only at the end of his project but without employment too.

One place that Hobhouse's sudden availability had not gone without notice was the Sociological Society. As we have seen, Hobhouse had been involved with the Society since its founding and had contributed to its activities in a variety of ways. However, on account of his time-consuming job at the *Tribune*, he had played little part in the Society's affairs over the course of the previous year. All that was about to change, though. In September 1907, Hobhouse was offered the Martin White chair of sociology at the London School of Economics – the first professional sociology job at a British university. Whilst Hobhouse's involvement with the Sociological Society since its inception, his publication of a substantive work of sociology in 1906, and his unemployment in early 1907 help explain how he came to be the first British professor of sociology later that year, they do not explain why the chair was offered to him, and not to anyone else. After all, for all his impassioned writings on the subject, Hobhouse, unlike his contemporary Patrick Geddes, had shown no great interest in establishing himself as a sociologist with institutional backing. As we will see over the course of the next two chapters, to understand why it was Hobhouse who became the Martin White Professor of Sociology at the LSE, we must chart what happened at the Sociological Society between 1903, the year that it was founded, and 1907, the year of Hobhouse's appointment.

Part III

Part II

5
The Origins and Growth
of the Sociological Society

In October 1904, Victor Branford, secretary of the recently formed Sociological Society, wrote to his friend and intellectual mentor Patrick Geddes. 'I did not make it clear in my previous letter', Branford told Geddes, 'that I had been seeing [Francis] Galton on the specific question of organising Eugenic research, either in connection with the Sociological Society or some other institution'. 'I took the line', Branford explained, 'that eugenic research might appropriately constitute a sectional department of a *sociological laboratory*, and that is how the Outlook Tower conception came in. If other conditions were such as commended themselves to [Galton]', Branford went on, 'I gather that he is prepared to endow with a considerable fund some form of eugenic research. He even contemplates going so far as to provide funds for securing the whole time of some trained biologist with an established reputation'.[1] As this chapter makes clear, these links between Branford, Geddes, Galton, and the Sociological Society were a defining feature of the early twentieth-century effort to establish a disciplinary framework for sociology in the UK. Indeed, for Branford, who took the leading role in coordinating those efforts, the aim was to make British sociology a Geddesian enterprise that was underpinned by Galtonian eugenics.

Given that it represented the start of a process that would end with sociology being established within the British academy, the formation of the Sociological Society is generally recognized as an important moment in the history of sociology in the UK. Somewhat surprisingly, though, there are very few close studies of the Sociological Society or of the culmination of its early activities: the appointment of L. T. Hobhouse as Martin White Professor of Sociology at the London School of Economics in 1907 and as Editor of the *Sociological Review* in 1908 – the first professional sociology posts of their kind in Britain.[2] To be sure, this lack of engagement stems in part from an apparent dearth of archival source material, which is most obvious in the cases of the Sociological Society, whose earliest records have long since vanished, and Geddes' papers, which suffer from the effects of a doomed trans-Atlantic voyage of 1915. In this respect, it is unsurprising that writing about the formative moments in early British sociology is either based entirely on published sources or characterized by far from compelling speculation.[3] However, notwithstanding the apparent lack of documentary evidence, the state of scholarship on early British Sociology can also be seen as a consequence of an intellectual consensus, drawn mainly from the work of R. J. Halliday and Philip Abrams, which has satisfied most historians of sociology, many of them sociologists themselves, since the late 1960s.[4]

According to received views, the Sociological Society was created in 1903 by supporters of Geddes who wanted to establish an institutional outlet for his work on sociology. To make the Society a reality, Geddes' supporters, in particular Victor Branford, built a coalition of themselves and two other groups: Galton's eugenicists and a loose collective of liberal and 'ethical' sociologists with whom Hobhouse was associated. Whilst these factions shared a general interest in the idea of a 'science of society' that could help guide social action, they also diverged from each other, the received view argues, on a number of significant points. Most importantly of all, each of the three groups that made up the Society had a different understanding of biological evolution and its social scientific applications – an issue that was seen as a prerequisite for the discussion of sociology. Thus, after a series of heated debates, which took place over the course of three and a half years, both the union at the heart of the Sociological Society and Geddes' dream of being crowned Britain's leading sociologist were shattered when the eugenicists broke away to form the Eugenics Education Society and Hobhouse was given the two most powerful jobs in the newly established field.[5] Consequently, Geddes was left isolated, the received view concludes, and unable to shape the agenda of the field his supporters had effectively created.

As with all received views, this story of early British sociology has some grounding in historical fact. For example, it is true that the initial push for a Sociological Society came from Geddes' supporters and that it was their intention for it to serve as a vehicle for his ideas. However, as recent work on Geddes, Branford, and Hobhouse by Steve Fuller, John Scott, Christopher Husbands, and Maggie Studholme has shown, the received view is largely inadequate as an account of the intellectual dynamics that drove the British sociology project forwards during its early years. Focusing on the question of what was at stake when Hobhouse was appointed at the LSE and the *Sociological Review*, Fuller has challenged Scott, Husbands, and Studholme to reconsider not only the nature of the relationship between biology and society in Geddes and Branford's thought but also how those links contributed to Geddes' failure to secure the first British chair of sociology.[6] According to Fuller, scholars such as Studholme, Scott, and Husbands are so deeply committed to seeing Geddes as a left-leaning reformist thinker that they have glossed over important but uncomfortable issues regarding the connection between biology and society in his work. Indeed, for Fuller, we need to take seriously the idea that the intellectual divide between Geddes and the eugenicists, with whom few of Geddes' current admirers would identify themselves, was nowhere near as wide as the received view has always suggested it was.

Whilst this recent debate has signalled the inadequacy of most existing interpretations of the early disciplinary history of British sociology, it has also highlighted the need for a more detailed understanding of what actually went on at the Sociological Society between 1903 and 1907. Building on the exploration in Chapter 1 of the late 1870s origins of the debate about sociology in the UK, as well as the reinterpretations of Galton, Geddes, and Hobhouse in Chapters 2, 3, and 4, this chapter and the next address that need by casting new light on the relationships between the key figures in early twentieth-century British sociology. Indeed, by utilizing previously unknown correspondence between the major players in the Sociological Society's early activities, this chapter provides the most complete account yet of the formative first 18 months of the organization's existence, a period about which so little has previously been known, and documents how disputes about the relationship between biology, sociology, and social action laid the foundations for Hobhousean, rather than Galtonian or Geddesian, sociology in the UK.

Like recent scholarship on the subject, this new account of British sociology's early institutional life involves recovering both the intellectual context in which it was established and significant but largely forgotten contributions to the development of the field. For instance, by highlighting Victor Branford's pivotal role in not only getting the British sociology project off of the ground but also keeping it together when profound disagreements threatened to break it up, this chapter and the next support John Scott and Christopher Husband's calls for Branford to be recognized as a founding father of the field. However, whilst confirming the significance that scholars have recently attributed to Branford's part in these early twentieth-century events, this chapter also introduces a new figure, Lady Victoria Welby (1837–1912), whose importance to the early history of British sociology has never before been documented.

A former maid of honour to Queen Victoria, Lady Welby had rejected court life and instead spent the late nineteenth century establishing her credentials as a philosopher. Whilst developing what she called 'significs', a philosophical system for analysing language that a number of scholars have seen as a foundation stone of the modern field of semiotics, Welby became a fixture at learned organizations, such as the Aristotelian Society, and a correspondent of leading thinkers from the natural and social sciences as well as philosophy.[7] From the late 1880s onwards, this network of correspondents included many of those who would go on to dominate British sociology in the early twentieth century, including Galton, Geddes, Hobhouse, and Branford. Through these connections, Welby became a figure of great importance at the Sociological Society and it was her ability to communicate with each of the groups who were involved that helped Branford keep the organization together at crucial moments in its early years. Indeed, without Welby, the trajectory of the Sociological Society would have been very different.

As we will see first, the impetus behind the founding of the Sociological Society came from Edinburgh, where Geddes' attempts to set himself up as a sociologist had, not for the first time, failed. Spurred on by a desire to make Geddes' vision of a 'Sociological Institute' a reality, Branford took control of the idea by transplanting it to London, where, he reasoned, there was a much greater chance of securing the support it had failed to attract in the Scottish capital. However, in so doing, Branford was confronted by a wider set of questions, which, we will see in the second section, shaped the early discussions about sociology in seldom-recognized ways. Driven by the memory of the late 1870s and early 1880s debates about the exclusion of excessive speculation from social science, Branford, with the help of Lady Welby, actively sought out Francis Galton who, as a veteran of those late nineteenth-century controversies, Branford saw as the ideal person to give British sociology a reputation for scientific rigour. It was in this sense that, as the third and fourth sections explain, Branford's intention was to unite Galton and Geddes through their shared vision of social science securing social progress by guiding the laws of evolution – an idea that Geddes himself approved of when he made his first appearance in front of the

Sociological Society. As we will see at the close of this chapter, though, it was the initial failure of Galton, Geddes, and the sociological project to gel, both intellectually and practically, which led the field down an alternative path that would lead eventually to Hobhouse.

From the Outlook Tower to the Sociological Society

'I managed to spend a couple of hours with [the geographer J. G.] Bartholomew last night and talk re[garding the Outlook] Tower', wrote Branford to Geddes in early July 1902. Not for the first time, Geddes' Outlook Tower project was experiencing financial difficulties and his friends were trying to think of ways to help save it. According to Branford, Bartholomew had one 'last suggestion', which was to 'form a Sociological Club or Association to support the Tower by its annual subscribers'. If such an organization was successful it might be possible, Branford suggested, to persuade someone to buy the building and then lease it back on favourable terms.[8] Inspired by this idea, Geddes decided to attempt to form a 'Scottish Society of Sociology', the realization of which, Bartholomew told Branford in March 1903, Geddes 'look[ed] to... in some sense as the attainment of his life's work so far'.[9] However, despite the best efforts of Geddes, Branford, Bartholomew, and others, the money that the project required was not forthcoming and, by mid-1903, Geddes and Branford had resigned themselves to the possibility that, if it was to become a reality, a society for sociology would have to be somewhere other than Edinburgh.

The timing of this push to establish Geddes as a social scientist with an institutional home of some kind was brought about by a number of factors. His sole permanent employment in 1903 was still the chair of botany at the University of Dundee – a job that had been created specially for him by his former student and long-term benefactor James Martin White. Whilst White's stipulation that Geddes be required to attend the university during the summer months only had left Geddes with plenty of free time, the arrangement was not without its drawbacks. Aside from the fact Geddes' limited duties meant his remuneration was actually fairly small, his failure to fully develop the Edinburgh School of Sociology that had captured his imagination in the opening years of the twentieth century or the institute that Bartholomew had suggested had left Geddes, as he confessed to his friend Lady Victoria Welby in April 1903, with the feeling of an 'increasing urgency of finding some work for my vacant 40 weeks'. To this end, Geddes had become increasingly interested in the prospect of becoming involved with social science in London; specifically with any openings there might be for 'teaching sociology' or, indeed, in the possible 'demand for a Sociological Review, which [he] could edit', he thought, 'without losing [the] interest of readers'.[10]

Although Geddes' ability to seize on opportunities of this kind was restricted in early 1903 by his commitment to a Carnegie-funded project to redevelop the Scottish city of Dunfermline, an episode that will be discussed in the fourth section of this chapter, Branford had, by late April, become 'very busy' corresponding with contacts in London.[11] In trying to interest people in the plan for an organization dedicated to sociology, Branford had made an important change to the approach that had been tried without success at the Outlook Tower. Rather than pursuing people on the basis of their support for Geddes, Branford had opened discussions with a broad range of people whom he thought would be interested in the general idea of developing sociology – a subject that had yet to gain formal recognition within British academia. Thus, whilst his ultimate aim was to create something that would benefit Geddes, Branford was attempting to sell the project as the first, non-partisan step towards embedding sociology within the UK's university system.

After the difficulties that had characterized Geddes' plans for the Outlook Tower, Branford's proposals for a sociology organization proved to be an immediate success. Tapping into groups such as the Charity Organisation Society (COS), which aimed to bring voluntary organizations under a single philosophy of reform and was headed by the social reformer Charles Loch, as well as the large numbers of historians, philosophers, psychologists, and politicians who could be found in and around the London area, Branford organized meetings in May and June 1903 to 'consider whether the opportunity was suitable for the formation of a Society to promote sociological studies'.[12] At the second of these meetings, which were held on the premises of the Royal Statistical Society, over 50 delegates, including Hobhouse, Geddes, White, and the anthropologist A. C. Haddon, a friend of Geddes who had participated in the summer schools at the Outlook Tower, resoundingly approved Branford's proposals and resolved to push the project forwards. However, whilst this endorsement was subsequently presented as an act of spontaneous enthusiasm by the Sociological Society, Branford, assisted by a small provisional committee, had actually carefully stage-managed the event.[13] Indeed, in addition to organizing all of the speeches in advance, Branford and the committee had even informed delegates of the arrangements for forming a 'Sociological Society' immediately after the meeting had closed.[14]

Stage-managed or not, the support meant little without the money to substantiate it and, in this sense, Branford's new Sociological Society was also a success in ways that had always eluded the Outlook Tower projects. After covering all of the Society's early costs, White, who had inherited a substantial estate from his father in 1884, quickly indicated he was willing to bankroll the schemes for promoting sociology that had been discussed at the meetings of May and June.[15] Writing to Sir Arthur Rücker, the principal of the University of London, on the day in late June when the Sociological Society was formed, White offered £1,000 'to be expended over several years in providing a Preliminary Course or Course of Lectures in Sociology'. By sociology, White explained that he meant 'the study of social organisation, development and ideals, past and present, over the world, from the lowest to the highest forms; with the object not only of constructing a scientific theory of society, but also of associating such theory with the highest philosophical thought, and of indicating the bearing of such knowledge on practical life'.[16]

After considering this offer, the Senate of the University of London, following a suggestion from White himself, decided to appoint a Committee, which included White, Haddon, Rücker, and the Liberal MP James Bryce, who had been involved in social reform projects such as Toynbee Hall, to examine the proposals.[17] Reporting back to the Senate in mid-November, the Martin White Benefaction Committee recommended that the University accept the offer but that it should do so only on certain conditions. Primarily, it had to be recognized, the Committee argued, that there were already 'a large number of courses' in London that were 'connected with the study of Sociology', such as those at the COS's recently established School of Sociology and Social Economics, which had been established in 1903 as a place where social workers and investigators could be trained in the COS's social philosophy of individual responsibility.[18] The Committee therefore stressed that, whilst they believed that the University should embrace sociology, great care had to be taken to ensure that what White and the Sociological Society were proposing was both novel and worthy of a place on the curriculum.[19] Although this stipulation allowed the Committee to exercise a certain amount of control over sociology within the University of London, it did not, as the familiar names on the Martin White Benefaction Committee demonstrated, remove the Society's influence over the development of sociology within academia. Indeed, as White reflected in his letter to Rücker, whilst there was no formal connection between the two, it was clear that 'the same people are interested both in the Society and in Sociological teaching'.[20]

What the University of London's response to White's offer also indicated, though, was the genuine uncertainty that existed with regard to the question of what exactly sociology was. The Sociological Society initially sought to combat these problems through a general statement of its purposes that was subsequently appended to the first collection of *Sociological Papers* – the edited volumes the Society issued during the first three years of its existence as a record of its meetings. Their aims were, the Society declared, 'scientific, educational, and practical' and they would achieve them by providing 'the common ground on which workers from all fields and schools may profitably meet – geographer and naturalist, anthropologist and archaeologist, historian and philologist, psychologist and moralist'. However, far from repeating what they did in their own disciplines, these investigators would 'all be contributing their results', the Society suggested, 'towards a fuller social philosophy'; one that included 'the natural and civil history of man, his achievements and his ideals'. In so doing, these investigators would bring into focus 'the conditions and forces which respectively hinder or help development, which make towards degeneration or towards progress' – something that they were unable to do when working in isolation from one another.[21] Indeed, as the liberal periodical the *Speaker* observed, the Sociological Society's call for minds to be focused on these goals reflected a belief that 'social evolution in industry, politics, and other spheres of conduct is largely modifiable by human reason and the general will'.[22]

Buoyed by these successes, Branford and the Society's Council began plotting their next moves and, during a series of meetings in late 1903 and early 1904, they settled on a strategy that was shaped by a concern to demonstrate clearly what sociological research looked like and what it might achieve in the future.[23] The Society would concentrate, the Council decided, on organizing meetings for the remainder of the 1903–4 academic year. At those meetings, presentations would be listened to, discussed, and then published in the *Sociological Papers*, which would be prepared by a committee consisting of the MPs J. M. Robertson and C. M. Douglas, the writers H. G. Wells and Benjamin Kidd, as well as Geddes and Hobhouse.[24] However, it was apparent to Branford that the Society's, and ultimately Geddes', aims would only be met if its meetings and journal were of a high quality. To this end, Branford began issuing invitations to the people from whom he wanted the Society to hear. In so doing, his attention became fixed on Francis Galton – the one figure Branford thought was capable of underlining the Sociological Society's scientific credentials. Having not only led the campaign, explored in Chapter 1, to close Section F, the social science branch of the British Association for the Advancement of Science, some 25 years earlier but also developed a heavily statistical form of social investigation that had captured the imaginations of many biologists, Galton stood for high scientific ideals. For this reason, Branford was determined to get Galton involved with proceedings at the Sociological Society.

Sociology and the eugenicists

After nearly four decades of trying to convince the scientific community and general public that the key to social progress was a deeper understanding of the laws of inheritance, Galton found Britain to be far more receptive to his ideas in the early twentieth century than it had been during the late nineteenth. Whilst Karl Pearson, W. F. R. Weldon, and William Bateson, in particular, were championing his work in the scientific community, political events, such as the outcry over the physical fitness of army volunteers during the Boer War, had also helped create an audience eager to hear more about eugenics.[25] As a consequence, at the start of the twentieth century, Galton had renewed his campaign for eugenics to be taken up as a national concern. In this sense, his call for society to be scientific about itself resonated clearly with the ideas coming out of the Sociological Society in late 1903. However, despite donating £5 to its costs and writing what Branford described as 'a very encouraging letter' about its potential, Galton was initially unwilling to be closely associated with the Society.[26] In part, this disinclination was due to the restrictions age had placed on the 81 year-old Galton's activities. Yet, as Branford set about persuading him to take a more visible role in the Society's affairs, it became clear not only that Galton and his associates were troubled by the nature of the Society's activities but also that many of Branford's aspirations for the Society were dependent on satisfying those concerns.

As he told Lady Welby, who first approached Galton about the Sociological Society on his behalf, in January 1904, Branford believed Galton's letter of the previous year was the Society's 'most valuable asset'.[27] This enthusiasm for Galton's apparent 'approval [of] and sympathy' with the Society's aims was rooted in the fact Branford saw in Galton something he did not in the economists, psychologists, philosophers, anthropologists, and politicians who were already on board at the Society: a direct route to scientific respectability.[28] For Branford and others, the controversial 1870s debates about political economy, which were explored in Chapter 1, still cast a long shadow over the social sciences in Britain and, as a consequence, there were concerns at the Sociological Society about their potential to repeat the kinds of rabble-rousing discussions that had almost brought an end to the BAAS' Section F in 1878 and then subsequently contributed to the demise of the National Association for the Promotion of Social Science in the mid-1880s.[29] As a key figure in those late 1870s debates at the BAAS, Galton, winner of the Royal Society's Darwin Medal in 1902, was exactly the kind of person Branford wanted involved in the British sociology project.

Galton, however, initially rebuffed Branford's requests for him to take a prominent role in the Sociological Society's affairs. After Branford had suggested to Galton that the Society wanted him to contribute to a meeting on eugenics, Galton first responded with the offer of a paper just ten minutes in length and a recommendation the Society might think about 'the idea of a volume of essays by different writers, hoping – perhaps hopelessly – that their writings will be less concerned with speculations than with conclusions from well ordered facts'.[30] Then, however, Galton reconsidered this proposal and decided he could not guarantee he would even attend, let alone present a paper, at such a meeting. For Branford and the Sociological Society, though, it was crucial Galton not only approve of what they were doing but be seen to approve of it as well, which made the notion of a session on eugenics without Galton, as Branford and Karl Pearson agreed, 'rather like the play of Hamlet with the prince left out'.[31]

Of course, Branford was also aware that Galton's level of involvement had serious consequences for the kind of cooperation that the Sociological Society could expect from the likes of Pearson, Weldon, and Bateson. Writing to Galton in April 1904, Branford explained that Pearson had agreed to chair a session on eugenics but only 'if [Galton] expressed a desire for him to preside'. Indeed, Pearson had 'made this promise', Branford went on, 'notwithstanding a certain want of sympathy with the Sociological Society' – an attitude that was rooted, Branford suggested, in a scepticism of whether 'the main objects of the Society will be the strictly scientific investigation of social phenomena'.[32] When pressed on this subject by Welby on Branford's behalf, Galton confessed that he found 'Pearson's reluctance...quite intelligible'. 'If the Sociological Society could get *solid* work done, and not merely speculative', Galton wrote, Pearson 'would warmly welcome it'. However, Galton concluded, 'the fiasco of the Social Science Association and the dearth of men engaged *intelligently* on *solid* work, fully justifies [Pearson's] hesitation'.[33]

Despite this scepticism, Galton was interested in what the Sociological Society had to offer. After hearing from Lady Welby about Branford's 'great desire to secure [his] interest and that of Prof. K. Pearson...in order effectually to prevent the very thing [they] both dread[ed] and deprecate[d]', Galton visited Branford at the London offices of his accountancy firm, which Branford had also been using for Sociological Society business, to resolve the stand off between them.[34] According to Branford, Galton's appearance on the 11 April 1904, just one week before the Society's programme of meetings was due to commence, resulted in Galton carrying out 'a searching cross-examination' of his ambitions for the organization.[35] Emerging from that discussion 'favourably impressed' by what Branford had to say, Galton agreed to give a full-length presentation the following month – something that would have left Branford feeling he had taken a giant step towards launching the Society and sociology more generally into the British scientific imagination.[36]

Given these behind-the-scenes manoeuvrings, it was no surprise that when the Sociological Society met for its inaugural session on the 18 April at the London School of Economics, which had agreed to lend its premises for event, the only person mentioned by name in the opening address of the Society's president James Bryce was Galton.[37] Billed as 'one of the most original and most unwearied of our investigators', Galton and his work was held up by Bryce as an example of the kind of progressive social science the Society wanted to encourage.[38] Thus, as Branford was finally able to confirm that Pearson had agreed to take the chair at the eugenics session, Galton would have been hopeful that his audience at the Sociological Society would be ready to both engage with his work and elevate his ideas about the role of rational reproduction in social progress to the top of British sociology's agenda.

In his paper, 'Eugenics: Its Definition, Scope and Aims', which has since come to be considered one of the most important statements in the history of eugenics, Galton talked generally about the theory and practice of eugenics, as well as how the Sociological Society could contribute towards making it a reality.[39] Eugenics was, Galton explained, 'the science which deals with all influences that improve the inborn qualities of a race' as well as 'those that develop them to the utmost advantage'.[40] However, it was important, Galton argued, that this definition not be taken to mean that eugenics was the breeding of all humans to a single archetype. 'There are a vast number of conflicting ideals, of alternative characters, of incompatible civilisations', he told the Sociological Society, 'but they are wanted to give fullness and interest to life'. After all, 'society would be very dull', Galton reasoned, 'if every man resembled the highly estimable Marcus Aurelius or Adam Bede'. The point of eugenics was therefore to create the conditions in which people could realize a variety of different capacities, from the artistic and the scientific to the practical and the theoretical. With people being educated about the laws governing reproduction, a eugenically aware society would then leave those people, Galton asserted, 'to work out their common civilisation in their own way'.[41]

The part Galton expected the Sociological Society to play in the realization of this eugenic future was closely tied to a three-stage process of development that he believed it was necessary for eugenics to go through if it was to achieve theoretical and practical maturity. It was important firstly, he told the Society, to make eugenics an 'academic question' because it was only through such means that its 'exact importance [would be] understood and accepted as a fact'. The reason, Galton went on, was that the second phase of development, during which eugenics would be 'recognised as a subject whose practical development deserves serious consideration', was dependent on sufficient scientific and technical knowledge of inheritance being acquired first. Only then could eugenics 'be introduced into the national conscience, like a new religion', he concluded, because the relative parity between the enthusiasm for his eugenic vision and the practical means of achieving it would be such that people could be expected to take on the reproductive responsibility it entailed.[42]

What the Sociological Society could do to assist this process, Galton suggested, was help collect and interpret the data on which stage one of the eugenics enterprise rested. More specifically, he believed that the Society should devote its resources to the compilation of information about 'thriving' families; that is, those in which 'the children have gained distinctly superior positions to those who were their class-mates in early life'.[43] Indeed, in a turn of phrase that was no doubt directed towards the Sociological Society's desire for scientific respectability, Galton told his audience that whilst the 'golden book' of thriving families would be a significant material contribution to the eugenic cause, it would 'have the further advantage of familiarising the public with the fact that Eugenics had at length become a subject of serious scientific study by an energetic Society'.[44] Thus, as he brought his presentation to a close and prepared for questions, Galton would have felt he had shown the Sociological Society exactly what it would take to achieve both its immediate needs and his further cooperation.

After all the effort that had gone into making this scenario a reality, there was a great deal riding on the Society's response to Galton's proposals. Indeed, as Branford knew only too well, Galton's already advanced support in the biological and statistical communities, as well as his huge personal wealth, meant the Sociological Society needed Galton a great deal more than he needed it. For this reason, it was the reaction of those present at the Society's eugenics meeting that provided the first major obstacle to the execution of the plan Branford had hatched almost one year earlier.

The fallout from the eugenics meeting

As Galton and Branford surveyed the audience gathered at the meeting of the Sociological Society on the 16 May 1904, they would have been quietly confident the session was going to be a success. From Galton's point of view, the audience featured enough of his close supporters, including Pearson and Weldon, as well as sympathizers, such as H. G. Wells and George Bernard Shaw, to make Branford's assurances about the seriousness of the Society seem justified. Moreover, from Branford's perspective, not only were there enough of Galton's most eminent supporters present to make a good discussion seem likely, but the numerous written statements he had received, including one from Bateson, seemed to convey the idea that the interest in proceedings went well beyond the room at the LSE. Yet as Pearson, taking advantage of his position as chair, rose first to speak, it quickly became clear that Branford's careful stage-managing was not going to bring about the event that either he or Galton wanted.

Declaring himself 'sceptical as to [the Sociological Society's] power to do effective work', Pearson used his opening comments to argue that it was impossible for a group of men and women, like those at the Society, simply to create 'a new branch of science'. What the Sociological Society needed, he asserted, was a single 'great thinker' who, like 'a Descartes, a Newton, a Virchow, a Darwin or a Pasteur', would blaze a scientific trail for their disciples to follow. Until that person had been found, 'a Sociological Society ... [was] a herd without its leader', Pearson argued, because 'there is no authority to set bounds to your science or to prescribe its functions'.[45] Although he no doubt thought Galton could be that leader, Pearson, who had clearly not taken leave of the scepticism that Branford had reported in his letters to Galton before the meeting, stopped well short of telling the Society that they would succeed if they rallied behind Galton's plans.[46]

If such sentiments were a bad start from Branford's point of view, the discussion that followed was even worse from Galton's perspective. One after another, contributors offered assessments of his paper that were cautious at best. Taking aim first, Henry Maudsley, who was known for his work on heredity, degeneration, and mental illness, put it to Galton that there were so many difficulties with the scientific understanding of heredity that 'we must not be hasty in coming to conclusions and laying down any rules for the breeding of human beings and the development of a Eugenic conscience'. The biggest problem, Maudsley argued, with the support of his fellow medical men, Dr. Mercier and Dr. Francis Warner, was the massive amount of variation amongst humans, which made it impossible to give an accurate prediction of what a specific couple's offspring would be like. To be that precise it would be necessary, Maudsley asserted, for the investigation of heredity to go 'far deeper down than we have been able to go...– to the germ-composing corpuscles, atoms, electrons, or whatever else there may be'.[47] Recognizing that Galton's work since the pangenesis debate of the early 1870s had been driven by a need to exclude just those kinds of considerations, Weldon jumped to Galton's defence when he told the audience that, 'as I conceive the matter, the essential object of Eugenics is not to put forward any theory of the causation of hereditary phenomena, but to obtain and diffuse a knowledge of what those phenomena really are'. As Weldon explained it to Maudsley and his supporters, Galton's 'actuarial method' operated at a level way above the kinds of individual concerns to which they referred. What Galton's work described was 'a large number of complex phenomena with a very fair degree of accuracy' and, 'for this reason', Weldon told the audience, Galton's methods were 'admirably suited for the purposes of eugenics'.[48]

Weldon's intervention did little, though, to stem the flow of criticism that was heading Galton's way. Whilst many contributors to the discussion raised the long-standing issue of whether it was nature or nurture that dominated inheritance, even those who saw rational reproduction as a good idea were minded to be sceptical of Galton's proposals. H. G. Wells, for example, argued that although Galton's paper featured a number of significant points, such as its emphasis on the existence of different types, rather than a single ideal, of human being, these ideas were 'only the beginning of a very big descent of concession, that may finally carry him very deep indeed'.[49] Furthermore, Alice Drysdale-Vickery, a physician and former member of the Malthusian League who was known for her advocacy of contraception as tool of female emancipation, echoed the views of others, such as Lady Welby and Francis

Warner, when she observed that 'heredity, as we study it at present, is very much a question of masculine heredity only'. If there was to be any hope of the kind of improvements of which Galton had spoken, Drysdale-Vickery argued, then Galton's account would have to recognize the role of women, not just men, in the process of inheritance so that 'women's specialised powers [could] be utilised for the intellectual advancement of the race'.[50]

Speaking last, Hobhouse, who, as we saw in the previous chapter, had recently published a series of stinging attacks on those who reduced social science to biology, managed to sum up the general feeling of the audience when he observed that 'it seem[ed] quite clear that before' heredity could be regulated, there needed to be a 'highly perfected knowledge' of how it worked. 'No one can doubt', he went on,

> that, if the kind of precise knowledge which I desiderate could be laid before us by the biologist, it would have considerable influence on our views not only of what is ethically right, but of what could be legislatively enforced... [After all,] the conception of conscious selection as a way in which educated society would deal with stock is infinitely higher than that of natural selection with which biologists have confronted every proposal of sociology... But until we have far more knowledge and agreement as to the criteria of conscious selection, I fear we cannot, as sociologists, expect to do much for society on these lines.[51]

Responding to this critical reception, Galton made it clear that he was furious with his audience's response. 'Much of what had been said might have been appropriately urged forty years ago, before accurate measurement of the statistical effects of heredity had been commenced', he declared, 'but it was quite obsolete' at the start of the twentieth century.[52] Choosing to say little more at the meeting itself, Galton instead took his frustration to Branford, who was forced to battle to convince Galton that the entire evening had not been a waste of time. Writing to him the day after the meeting, Branford apologized to Galton 'for the failure to bring forward a larger number of statisticians & biologists' to hear his paper. However, this failure 'was due to no lack of intention', Branford wrote, 'but entirely to the inability to attend, of many of those invited to take part in the Discussion'. Indeed, in an attempt to reassure Galton, Branford emphasized that 'the letters of declinature showed, for the most part, the warmest enthusiasm for the subject & expressed an obviously sincere regret of the inability of the writers to personally attend or contribute'.[53] As he confessed in a letter to Lady Welby, though, Branford did feel he had to take some responsibility for 'not having made sure of securing the presence and participation of a larger number of men whom Galton [had] already indoctrinated with eugenic conceptions...For this', Branford went on, 'I relied too much on the chairman', Karl Pearson, who had made his disdain for the whole enterprise clear for all to see. In this sense, Branford had to admit that he 'deeply sympathise[d] with [Galton's] disappointment over the discussion'.[54]

'Looking at [the meeting] in a historical way', though, Branford believed the way Galton's ideas had apparently fallen on deaf ears would be 'confirmatory of the magnitude of the occasion' at the Sociological Society. 'To go no further back than a couple of generations', Branford told Lady Welby, 'there is the analogy of Helmholtz's paper to the Berlin Academy of Sciences, of Joule's paper to the British Association and of Jevons' paper to Sect. F. of the British Assoc[iation]. In each case a new branch of science resulted [even] though the authoritative specialists who took part in the discussion were antagonistic'. For this reason, Branford hoped Galton would not let his first experience of the Sociological Society discourage him from being involved with its future. However, despite 'having said so much by way of trying to explain (away) the poverty of the discussion', Branford told Lady Welby it also had to be admitted that there was more to explaining Galton's 'pessimism about the Society' than what had happened after his presentation. 'For Galton, as a personality constituting one of the great spiritual forces of the time', Branford wrote,

> I have I hope a deep appreciation, and for Galton as one of the rarest of all initiators – the inventor of a new scientific method – I have the admiration which the historical student of methodology must necessarily feel. But I cannot for a moment admit that Galton has any status as a critic of the future of the General Science of Sociology. For there is no evidence to show that he has any adequate acquaintance with its past. I don't think he has any considerable acquaintance with the tradition of generalising sociological thought grown up through the labours of Vico, Herder, Condorcet and Comte. In a word he does not know enough of the history of sociology to predict its future, either in the hands of the Sociological Society or any other body. Hence his criticism comes dangerously near being the analogue of that which Maudsley and Mercier bring against Eugenics.[55]

Thus, as much as he wanted Galton to be involved with the development of sociology in Britain, Branford was not minded to surrender the field to him. Indeed, it seemed to Branford that the only 'proper line of development at present in regard to Eugenics is to take up the subject of "the Golden Book of Thriving Families"', which would be a relatively uncontroversial exercise that would keep Galton interested in the Society whilst it continued to carve out a place on the British intellectual landscape.[56]

The reason Branford had no intention of giving Galton a free rein over the Sociological Society was, of course, Branford's intention for eugenics to be the scientifically rigorous platform on which sociology would be built as a Geddesian enterprise. With respect to the fact they had both produced biosocial programmes that were designed to guide social action, Galton and Geddes shared a great deal of common ground. However, it was not immediately obvious to everyone how Galton's talk of guiding the laws of heredity was compatible with Geddes' ideas about civic improvement and systematizing the social sciences under the heading of sociology. To this end, Branford had arranged for the next session at the Sociological Society to be a theoretically orientated one, at which papers entitled 'On the Relation of Sociology to the Social Sciences and to Philosophy' would be heard from himself and the prominent French sociologist Emile Durkheim. Branford's thinking was that by introducing such questions to the Society for the first time it would clear the ground for Geddes to present his own style of sociology and highlight how it could work with the programme Galton had already outlined.

However, in keeping with its function as a bridge between Galton and Geddes, the Branford/Durkheim meeting, which was held at the LSE on the 20 June 1904, had very little intellectual coherence of its own. Whilst both Branford and Durkheim, who was unable to attend the meeting and had his paper read for him by the British idealist philosopher Bernard Bosanquet, argued that sociology was a synthetic science, dealing with natural laws, which expressed the interdependence of the phenomena studied by the specialist social sciences, they disagreed, as Branford put it in a letter to Lady Welby, on the question 'of putting in the foreground, the scientific treatment of practical questions rather than "first principles", methodology &tc', with Branford insisting that theory had to come before practice.[57] Yet on account of the large numbers of written contributions to the meeting that Branford had managed to solicit, including statements from the French philosopher and anthropologist Lucien Levy-Bruhl and the German philosopher and sociologist Ferdinand Tönnies, there was no substantive discussion of any of the points that either Branford or Durkheim had made.[58] In fact, with over thirty, often lengthy, written communications being read at the meeting, there was no debate of any kind as one person after another had their say on the methodology of sociology.

This lack of discussion and coherence was, however, exactly what Branford had wanted. The meeting had been brimming with ideas and, as a consequence of its cosmopolitan cast of contributors, it had given the impression Britain was in the process of joining an international debate that was already well underway. Indeed, because of its intellectually chaotic form, the session had given the impression sociology was a burgeoning field waiting for someone to stamp their authority on it. The stage was therefore set for Geddes to make a grand intellectual entrance at the Sociological Society and, after securing for himself the last slot of the academic year, he began preparing himself for the first of what he envisaged as a series of two papers on the ideas he had been developing since the late 1870s.

Geddes and the case for civics as sociology

By mid-July 1904, when Geddes was putting the finishing touches to his first Sociological Society presentation, the effort to establish sociology as both a subject of public debate in Britain and a part of the University of London's teaching curriculum was well under way. In addition to the Society's meetings, the first set of Martin White-endowed sociology lectures, which were delivered by Geddes, Haddon, Hobhouse, and the Finnish anthropologist Edward Westermarck during the Lent and Easter terms, were about to draw to a close. However, aside from those lectures, which were entitled 'Cities and their Cultural Resources', Geddes had played a relatively minor role in the Sociological Society's public affairs. The major reason for Geddes' absence from the Society was, as mentioned earlier, his involvement with a project funded by the American philanthropist Andrew Carnegie to redevelop Dunfermline, located only a few miles north of Edinburgh, on the opposite side of the Firth of Forth. To find the best way of spending the $500,000 Carnegie had donated to the cause, a board of trustees had commissioned planning reports in mid-1903 from both Geddes and the architect T. H. Mawson. Aware that the trustees intended to choose just one of these reports, Geddes had therefore been busy throughout much of the Sociological Society's first year writing his report, which he hoped would convince the Carnegie trustees to put civics, the name he had given to his study of cities and his practical prescriptions for improving them, into practice in Dunfermline.

As he told Lady Welby in November 1903, Geddes had been asked by the trustees 'to report, not only on the laying out of the grounds [in the heart of Dunfermline], but "on the structures which should be erected in and around [the area]"'. Whilst what the trustees seem to have wanted was a design for a new park and buildings surrounding it, Geddes had made a characteristically broad reading of their commission. Indeed, as he told Lady Welby, he had decided that, 'as you cannot erect structures without having decided on *uses*', the trustees' request entailed 'all questions of civics, in fact almost the whole policy of culture'.[59] It was therefore no surprise that, after conducting a thorough photographic survey of Dunfermline, Geddes found himself 'constantly struggling' in January 1904 'with the encyclopaedic "middens" of landscape-gardening, museum construction, and social betterment' that his 'still far from finished report to the Carnegie Trust' required.[60] Believing it was necessary for each of his suggestions to be framed by a far-reaching explanation of its place in the wider scheme of his thought, Geddes wanted his report to be both a plan for Dunfermline and a systematic treatise on the principles of urban development. However, as J. H. Whitehouse, secretary of the Carnegie Dunfermline Trust, reported to Geddes in February 1904, there was 'a feeling with many' of the trustees that Geddes' report had become 'too long and [dealt] with matters outside [of his] commission'.[61] Indeed, when Geddes had finally submitted his brilliant but elaborate report, the board of trustees decided that, despite Geddes' pleas to the contrary, it was far too expensive to be implemented.

Undeterred by the lukewarm response he was receiving from the Carnegie Dunfermline Trust, Geddes attempted to build on the work he had done in late 1903 and early 1904 by publishing his report as a book, which he called *City Development: A Study of Parks, Gardens, and Culture-Institutes*.[62] Laying the first copies of *City Development* on a table in front of him at the meeting of the Sociological Society at the LSE on the 18 July 1904, two months after Galton had outlined why sociologists should subsume their activities in his eugenic campaign, Geddes hoped his experience in Dunfermline would aid his explanation of the biosocial urban sociology that had inspired his suggestions for remodelling the Scottish city. Using his paper, which was entitled 'Civics: As Applied Sociology', as an opportunity to elaborate on the general theory that framed *City Development*, Geddes went to the Sociological Society with the intention of explaining to his audience, which included Ebenezer

Howard, the founder of the Garden Cities movement, and Charles Booth, famed author of *Life and Labour of the People of London*, the principles that underpinned the sociological vision he had been developing since the late 1870s.

At the heart of Geddes' presentation to the Sociological Society was the notion that cities should not be seen as isolated entities but as the products of a much wider and evolving geographic region. Describing to his audience an imaginary journey that followed a river system throughout a valley region, Geddes explained how the 'panoramic view of a definite geographic region' could help the social investigator detect the links between the state of a city at any particular moment in time and the development of civilization as whole.[63] By taking a step back from the 'vast labyrinth' of a major city such as London, it was possible, Geddes suggested, to observe how cultural, economic, and social shifts, which took place on a much larger scale, helped shape the urban environment. As society changed, he argued, each new generation modified its environment in a way that said something important about how its people had lived their lives. For example, whilst people had built castles, markets, and churches at specific points in the past, they had most recently transformed the physical appearance of both cities and the countryside with factories and industrial-age transport systems. In this sense, Geddes concluded, the city was both a 'place in space' and a 'drama in time', which was full of the 'traces of all the past phases of evolution'.[64]

Approaching the city in this way also had the advantage, Geddes told his audience, of enabling the social scientist to grasp how people both shape and are shaped by their environments. More specifically, by looking at institutions, it was possible, he suggested, for the sociologist and the social surveyor to study how people came into contact with the ideas that dominated their phase of social evolution. For instance, on examining the subjects that were taught in schools, such as Latin, literature, religious instruction, and mathematics, it was possible, Geddes argued, to trace how children were moulded by the ideas that were impressed on them by society through their education. Indeed, 'the inordinate specialisation upon arithmetic [in schools], the exaggeration of all three R's', could be seen, Geddes reasoned, as 'plainly the survival of the demand for cheap yet efficient clerks, characteristic of the recent and contemporary financial period'.[65] Thus, he told the Sociological Society, both 'the city and its children...present a thoroughly parallel accumulation of survivals or recapitulations of the past in the present'.[66]

Although, due to the constraints of time, Geddes was unable to describe in any great depth the practical outcome of this analysis, he did emphasize that, for all his talk about the importance of history, his was a style of sociology that looked forwards as well as back. As he put it towards the end of his talk,

> having taken full note of places as they were and are, of things as they have come about, and of people as they are – of their occupations, families, and institutions, their ideas and ideals – may we not to some extent discern, then patiently plan out, at length boldly suggest, something of their actual or potential development?

However, in emphasizing how important this idea was to his work, Geddes broke from the script he had prepared and circulated in advance. Amongst the most significant of these digressions came after his introductory remarks, when he somewhat unexpectedly started to talk about the relationship between his work and that of Francis Galton. Echoing his 1890 review of Galton's book *Natural Inheritance*, Geddes described the form of eugenics that the Sociological Society had heard about two months earlier as the 'study of the community in the aggregate'; that is, a style of investigation that saw society as the sum of its individual parts. Yet, Geddes went on, eugenics found 'its natural parallel and complement in [his] study of the community as an integrate, with material and immaterial structures and functions, which we call the City'. In this sense, he explained, 'the improvement of the individuals of the community, which is the aim of eugenics, involves a corresponding civic progress'.[67]

Whilst this late addition of eugenics to Geddes' paper was no doubt related to the ongoing efforts to keep Galton involved with the Sociological Society, it was by no means out of sync with what Geddes said in the rest of his presentation. As Geddes reminded his audience, for all his recent work on cities, which was symbolized by the copies of *City Development* in front of him, he was still 'primarily a student of living nature in evolution' – a statement with which the liberal MP J. M. Robertson agreed when he described Geddes' paper as an 'application of the view of a biologist to Sociology'.[68] Moreover, in the little he did say about his prescriptions for a society in tune with the sociological ideas he had outlined, Geddes alluded to the point where his recommendations for the social integrate converged with Galton's for the social aggregate. 'Few types nowadays...keep strictly to their period', Geddes asserted,

> we are all more or less mixed and modernised. Still, whether by temporal or spiritual compulsion, whether for the sake of bread or

honour, each mainly and practically stands by his order, and acts
with the social formation he belongs to. Thus, now on the question
of... practical civics, that is, of the applied sociology, of each individual, each body of interests may be broadly defined; it is to emphasise
his particular historic type, his social formation and influence in the
civic whole, if not indeed to dominate this as far as may be.[69]

In this respect, Galton and Geddes' vision of a progressive future shared
a belief in the importance of maintaining the diversity of types that
evolution had brought about – a point of agreement that highlighted
why Branford, in a letter to Lady Welby, had described Geddes as a
'Galtonian', albeit 'a critical one'.[70]

Despite these efforts to tie what he had said to the more concrete
suggestions Galton had laid before the Sociological Society two months
earlier, there were indications not everyone was overwhelmed by what
they had heard from Geddes. As numerous members of his audience
commented, Geddes' presentation, from its failure to explain what part
the Sociological Society would play in his scheme, to his irregular use
of practical examples, felt incomplete. Indeed, even though Geddes
responded to his critics, such as Robertson, by pointing out that his
paper was 'but the first half of [his] subject', there was a sense that
Geddes had yet to do enough to earn the wholehearted support of those
who were not already converts. In particular, Geddes, despite his claim
that civics was a theoretical and practical enterprise, had done little to
appeal to those members of the Society who were primarily interested
in how ideas could be channelled into social reform. Reiterating the
Spencerian scepticism of government that had permeated his previous
writings on society and evolution, including *The Evolution of Sex*, Geddes
instead told the Sociological Society that 'representative government'
could not be counted on to affect social change because, in his view, it
had 'fail[ed] to yield all that its inventors hoped of it'.[71]

Although Geddes gave vague hints about 'the united forces of civic survey and civic service' finding some kind of 'middle course' for change,
it was clear that anyone who was not already highly attuned to his ideas
about the evolution of co-operative and altruistic behaviour, which had
found its clearest expression in his work on 'reciprocal accommodation',
would have had great difficulty grasping exactly how he saw social
progress taking place. Indeed, there was a growing feeling amongst his
friends that, after spending over a decade in Edinburgh surrounded
by disciples rather than critical supporters, Geddes had fallen, as Lady
Welby wrote to Branford just four days before the Sociological Society

meeting, into a 'deep though unconscious rut'.[72] Looking at her copy of *City Development*, Lady Welby, at whose house in Harrow Geddes had finished his original Dunfermline report in early 1904, felt the comments Geddes had asked for had put her 'in a difficulty'. Although Lady Welby was 'glad' Geddes had managed to publish his report, she was somewhat concerned by the direction Geddes was taking in his work. Specifically, she felt his growing taste for neologisms and idiosyncratic terms, a tendency that she called 'Culture Jargon', was putting his work at risk of descending into 'clichés and tags'.[73] If he was serious about his ambitions to break out of the natural sciences and into the worlds of sociology and town planning, Geddes 'really need[ed]', Lady Welby suggested, 'to get outside the groove into which [his] work [had] fallen, and come to it with a fresh eye'.[74]

Spurred on by the criticism he had received from Welby and others, Geddes, who turned 50 in 1904, began to reconsider what he would have to do if he was to capitalize on what Branford had already achieved with the Sociological Society in London. Whilst he had made an important gesture towards Galton, which gave Branford hope a practical and intellectual alliance between civics and eugenics might still be forged, Geddes still had a great deal to do if Branford's hopes for the Society were to be realized. Thus, in late August, Geddes began to seek the advice of his friends and supporters about what he should do next. According to Haddon, whilst it was clear Geddes should continue to work on civics, he needed to think closely about whether he could do so effectively when he was living in 'Edinburgh & London at the same time'.[75] A decision needed to be made, Haddon told Geddes, because it was one that would have serious implications for his future ambitions.

What added to the potential importance of Geddes' deliberations and Branford's plans for the future of the Sociological Society was the fact Galton had made it known he was about to make a substantial amount of money available to projects that supported his ambitions for eugenics. Aged 88 and, ironically for the founder of eugenics, childless, Galton wanted to use his massive personal fortune to establish eugenics within British universities and intellectual life. Perhaps surprisingly, given his presentation in May had not gone quite as either he or Branford had hoped, Galton was still open to the idea that the Sociological Society might be the recipient of at least some of his money. Yet, as Branford very well knew, if Galton decided not to put any of his vast fortune into the British sociology project, it would be a massive vote of no confidence in an endeavour he had initially lent his support to. In this sense, whilst the first year of the Society's existence might not have unfolded in the way Branford had expected, the situation was delicately balanced in mid-1904 because his ambitions for sociology to derive its authority from Galton but be led by Geddes in the UK was still on the cards. However, as we will now see, it was the resolution of that situation that led British sociology to take a very different direction over the course of the following three years when it opted instead for the profoundly anti-biological vision of L. T. Hobhouse.

6
The End of Biological Sociology in Britain

In January 1907, Victor Branford wrote to Lady Victoria Welby to ask if she was 'well enough to be troubled with a matter which, [he knew], would secure [her] interest'. With the Sociological Society approaching four years of age, Branford was deeply concerned by the fact that the UK's sociology movement was 'inclining to be dispersive'. Consequently, he had come up with a strategy to rejuvenate the field and restore its sense of purpose. However, Branford's plan, which has been unknown to historians of sociology until now, involved a dramatic departure from the path he had originally argued British sociology should take. According to Branford, it was Leonard Trelawny Hobhouse, rather than Patrick Geddes or Francis Galton, who was 'the one personality round which the whole [British sociological] movement...might be crystallized and concentrated'. Branford had therefore decided that the Sociological Society should push for the creation of a 'Sociological Faculty', one that would be 'unif[ied]...around a Chair or Lectureship of General Sociology for Hobhouse'.[1] As this chapter shows, Branford's plan, which led to Hobhouse, an avowedly anti-biological social thinker, holding the UK's first chair of sociology, was inspired by the Sociological Society's scepticism towards eugenics and civics – the two biosocial programmes Branford had initially wanted British sociologists to embrace.

As has recently been demonstrated by the exchange between Steve Fuller, on the one hand, and Maggie Studholme, John Scott, and Christopher Husbands, on the other, the reason Hobhouse was appointed professor of sociology at the London School of Economics in 1907 is still a matter of great debate.[2] Whilst opinions on this subject have often been inspired by what scholars have thought of Hobhouse's subsequent impact on British sociology, the focus for discussion has always been on one key question: why was the Martin White chair of sociology at the LSE awarded to Hobhouse instead of Geddes? After all, Geddes, unlike Hobhouse, had demonstrated a strong interest in sociology by frequently presenting his work to the Sociological Society, which had been created by Geddes' supporters. Moreover, the chair at the LSE was endowed by James Martin White who had not only bankrolled the Society but also created a professorship for Geddes at the University of Dundee in 1889. Given all of these advantages, it is often asked, how did Geddes lose out to someone who, according to at least one of his biographers, never actively pursued the job?[3]

The answer to this question has frequently been pinned down to a single event. Based on a brief but ambiguous statement from Branford during the late 1920s and a story that was subsequently told by Geddes' son, Arthur, to Geddes' biographer, Patrick Boardman, scholars have believed that, at some point during 1906 or 1907, Geddes, Hobhouse, and two other candidates were interviewed at the LSE.[4] However, according to Arthur Geddes, his father 'rushed up [to London]' for this interview and 'gave a very bad lecture beforehand', which led to him being passed over for the chair.[5] For Geddes' supporters, both during his lifetime and after, this story has been seen as a moment of massive counterfactual significance for both his career and the subsequent history of British sociology. If he had been better prepared for the interview or if the intellectual elite of London had been more open to his seemingly leftfield ideas, then he would have been given the job, Geddes' supporters have argued, and with it the opportunity to shape sociology according to his civics research programme. Furthermore, his supporters have claimed, if the LSE had chosen Geddes, British sociology would not only have been very different but also theoretically sophisticated in a way that it is often argued it was not under Hobhouse.

Yet, as this chapter reveals, there never was an interview for the Martin White chair of sociology. In fact, Hobhouse became Britain's first professor of sociology because Branford, one of Geddes' closest supporters, selected him for the job in the belief Hobhouse could unify what was becoming an increasingly fractured field.[6] What is argued about these facts is that they should provide us with new points of reference for navigating the debate about why British sociology became a Hobhousean, rather than a Geddesian, science. Whilst Branford's plan compels us to treat Hobhouse's appointment as a matter of design rather than historical accident, it also forces us to ask why Branford had come to see Hobhouse, rather than Geddes, as the thinker around whom British sociology might be unified. As this chapter shows, the answer to that question needs to be understood as the outcome of the debates about the scope and identity of sociology, which had commenced in Britain with the founding of the Sociological Society in 1903. Indeed, by tracing how disputes about the role of biological ideas in sociology provided the context in which the British sociological project began to stall in 1905 and 1906, this chapter shows that Hobhouse was chosen for the Martin White chair because he, unlike Geddes, saw sociology as a science that was free from the intrusions of biology.

Like the previous chapter, this new account of the lead up to Hobhouse's appointment is based on previously unknown correspondence between key members of the British sociology project. In conjunction with what we have already seen, this correspondence confirms beyond doubt the crucial roles played in the early history of British sociology by two figures who have been largely absent from earlier accounts: Branford, whose plans gave shape and direction to British sociology, and Lady Victoria Welby, whose contacts and private interventions helped lay the path on which British sociology travelled. As we will see in the first two sections of this chapter, Branford's original ambitions for British sociology, explored in the previous chapter, were thwarted during the 1904–5 academic year when Francis Galton withdrew from the Sociological Society, apparently after consulting Lady Welby, and Geddes failed for a second time to convince the Society that civics was a viable research programme. As the third section shows, these developments knocked British sociology off of the course it had been travelling during the previous two years. With no notion of what would replace the Geddesian and Galtonian biosocial ideas that had featured so prominently in its early meetings, the Sociological Society drifted and, as we will see in the final section of the chapter, it was in this context that Branford formulated his plan, revealed first to Lady Welby, to make Hobhouse the effective head of British sociology. A writer on politics and social reform who had not only been involved with the Society since its inception but who had also recently published a well received work of sociology, Hobhouse had many of the credentials that were required of a leader of British sociology. However, what drew Branford to Hobhouse more than anything else was his rejection of the ideas about biologizing social science that threatened to bring the field to breaking point.

The end of Galtonian sociology

By the end of June 1904, Victor Branford felt he was beginning to see the first fruits of his campaign to get Francis Galton involved in the debates at the Sociological Society. Indeed, Branford was particularly excited, he told Lady Victoria Welby, by the fact that 'the Editor of *Nature*, writing to [him] as Secretary of the [Sociological] Society', had asked him if he would either review a book for the journal or 'recommend some one [else] to do it'. What seemed significant about this request, Branford explained to Lady Welby, was that the editor of *Nature* had approached him in 'a way that was entirely impersonal', which Branford had concluded was 'tantamount to a recognition' that 'the Society... [had] status in these matters'. For this development, Branford believed the Sociological Society had 'entirely to thank the presence of Mr. Galton among us at the May meeting'. In this sense, Branford went on, it was 'encouraging that responsible people' seemed to be of the opinion the Society could 'live up to' the kinds of scientific ideals that the Galton meeting had been organized to promote.[7]

Inspired by this kind of attention, Branford decided it was more important than ever for the Sociological Society to take up the cause of the 'Golden Book of Thriving Families', a database of information about high-achieving families, which Galton had suggested was the most appropriate way for the Society to contribute to his eugenic project. To try and get the 'Golden Book' up and running, Branford had taken the opportunity, he told Lady Welby in August 1904, 'of renewing [his] attack on Martin White (our one and only Sociological millionaire – and a very tough one!)', who Branford hoped would bankroll the idea, just as he had done everything else in the Society's early history.[8] However, in addition to the attention it might attract from the scientific establishment, there was another reason Branford was so keen to act on Galton's suggestion. As part of the next stage of his eugenics campaign, the heirless Galton had made it known to selected individuals, including Branford, that he planned to make a significant amount of his substantial wealth available to projects he saw as converging with the aims he had outlined in his presentation to the Sociological Society the previous May. Seeing in that money the opportunity to secure both Galton's seal of approval and further funds that could be used to make Geddes the leader of British sociology, Branford had decided to renew his efforts to persuade Galton that eugenics and the Society had a great deal in common.

To Branford's disappointment, though, White was not convinced that the 'Golden Book', an undoubtedly expensive undertaking, was a wise investment, which left Branford with no substantive evidence that the Sociological Society was contributing to the cause of eugenics. However, as he revealed in a letter of October 1904, Galton considered Branford's case very closely and, 'after conversations with many, found [that] there were six alternative plans, all definite and reasonable, and indeed two more [or] less definite', that he could pursue. 'Taking all in all', though, the University of London, rather than the Sociological Society, seemed to Galton to be the best-equipped organization to further his cause. Indeed, as he told Branford, Galton had been to see Sir Arthur Rücker, the principal of the University of London, on the 10 October and, 'after a long talk[,] settled preliminaries, so far that [Galton had] then and there [written] out a general proposal which as a matter of "urgency" was brought before the "Academic Council" that...very afternoon'. Having had his offer accepted, he told Branford 'prematurely and *privately*', Galton had given £1,500 'to maintain a Research Fellowship in National Eugenics [at the University] for three years with [the] expectation of renewal'.[9]

Alarmed by this development, Branford immediately wrote to Geddes to pass on Galton's news. 'You will see [from his letter]', Branford told Geddes, that 'there is every chance of this new endowment passing the Sociological Society (and your sociological interests which I think of as intimately correlated with the Society) without benefiting them in any way, except most remotely and indirectly'. Indeed, Branford observed, it looked as though Galton's money might pass Geddes by 'just as the White Endowment has done'. What made this development especially frustrating for Branford was the fact that Galton's decision had been driven by practical, rather than intellectual, concerns, which had apparently been suggested to him by one of the very people Branford had called on for support. 'I think in all probability', Branford told Geddes, that 'Galton would have given the endowment to the Sociological Society if we could have persuaded him that it was assured of continuity and permanence'. However, Branford went on, 'Lady Welby unfortunately put it in his head that I was overworked and he feared I was going to collapse and the Society with me'. As a consequence, 'the utmost' he hoped to 'achieve in connection with this Galton business' was 'an annual lecture...given by the Fellow appointed as a lecturer under the auspices of the Sociological Society', which could be 'published in [the Society's] Proceedings'.[10]

Despite Branford's sense of resignation, Geddes was not quite ready to give up on the effort to bring eugenics and the British sociology project together. In reply to his letter, Geddes urged Branford to consider whether there was still a chance for the Sociological Society to either hijack Galton's plans or, at the very least, to derive some benefit from them. 'Have you written to [J.] Arthur Thomson?', Geddes enquired of Branford. 'If not...don't you think you should write to [Thomson] strongly', Geddes argued, 'and try to get him to see that he, with his vast knowledge and his note of moral idealism, might make [eugenics] peculiarly his own?' Indeed, according to Geddes, a long-standing critic of Galton's approach to eugenics, it was Thomson, then Professor of Zoology at the University of Aberdeen, 'who should be the heir of Galton...not dry statisticians...like Pearson...[and] Weldon'. Whilst it would not be easy to change Galton's mind on these subjects, Geddes told Branford, doing so was 'the best opportunity of our lives alike of individual and co-operative usefulness'.[11]

In a bid to advance this plan, Geddes wrote to Lady Welby, whose counsel Branford believed had turned Galton away from the British sociology project, to ask if she could help reconnect sociology and eugenics by encouraging Galton to reconsider his attitude towards the Sociological Society. Although he possessed no real evidence to substantiate the claim, Geddes assured Lady Welby that the Society was in much better shape and was capable of much greater things than Galton had perceived it to be. Indeed, if Galton included the Sociological Society in his plans, Geddes told Lady Welby, then Branford 'would no doubt seek to combine the imagination and critical outlooks, as of [H. G.] Wells and [George Bernard] Shaw, the more biological approach of Thomson or myself, with the statistical work in which Pearson, Weldon, etc. are so active in carrying on the impulse of Mr. Galton'. However, Geddes went on, the University of London would 'be too apt to content itself with only the last named of these', which would exclude a whole take on eugenics that, in his view, was essential for its development.[12]

Yet, as Branford told Geddes in late October, these pleas were far too late. Galton had already gone ahead and 'given the whole of his endowment to the University of London without any reference whatever to the [Sociological] Society'.[13] The only thing that Branford believed was left for the Society to do was to hope the University of London might see some way to link eugenics and sociology. To this end, Branford wrote to the University's Academic Council in late October to inform them that,

> at a meeting of the Council of the Sociological Society on October 27th, the announcement of the University's acceptance of Mr Galton's endowment, was brought before the Council, and a resolution was adopted recording the gratification with which the Council heard of an event so well calculated to promote sociological studies; and at the same time expressing the willingness and desire of the Sociological Society to assist in any way...the research work undertaken under Mr.

Galton's endowment, should the University be desirous of accepting such assistance as the Sociological Society might be able to render.[14]

However, with respect to his grand plans for Galton to be directly involved with the Sociological Society, helping to shape its agenda and allowing it to trade off of his scientific reputation, Branford was resigned to the fact that he had failed. As if to underline this failure, Galton used the occasion of his next, and final, appearance at a Sociological Society meeting to give two presentations addressing the future of the eugenic movement. Without a hint of irony, Galton spoke first at the London School of Economics on Valentine's Day 1905 on the subject of 'Restrictions in Marriage'. Surveying the religious, cultural, and legal rules governing marriage around the world, he argued for the reworking of customs in light of eugenic ideals.[15] Then, in a paper entitled 'Studies in National Eugenics', which included a plan for a 'Biographical Index to Gifted Families', which sounded a great deal like the 'Golden Book of Thriving Families', Galton outlined the issues he considered appropriate subjects of investigation for Edgar Schuster – a former student of W. F. R. Weldon who had recently been appointed the first Francis Galton Research Fellow in National Eugenics.[16] On this occasion, though, none of Galton's most prominent scientific supporters were in attendance. Moreover, and perhaps as a consequence, there was none of the controversial debate that had occurred after his first paper to the Sociological Society.[17]

From Branford's perspective, however, there was one important, if somewhat belated, knock-on effect of Galton's turn away from the Sociological Society. After hearing in October 1904 of Galton's decision to fund the Fellowship in National Eugenics, Geddes had been forced to think again about the reasons behind his unsuccessful presentation to the Society the previous July. Admitting that the Outlook Tower, which had still to attract a major audience or financial backer, was a 'failure', Geddes told Lady Welby he was finally 'being compelled reluctantly to give up both [his] home … and work in Edinburgh'. Whilst his 'Summer Term's teaching in Dundee' was 'not only a flower-rich holiday and a main breadwinning resource, but [also] … an increasingly efficient piece of work', it was 'no adequate outlet' for his wider interests. As a consequence, Geddes was 'being reluctantly forced to think more seriously of making London [his and his wife's] head-quarters' and he was planning to use his lecturing engagements during the upcoming Lent term as a test run for the relocation. He was still unsure, though, about the likelihood of this move being a success. He did not think his lecturing during early 1904, both for the Martin White Benefaction and a range of extra-mural schemes, had gone particularly well, 'while the destruction of nature and atmosphere alike [in London], makes each day's work in it depressing beyond all the places I have ... tried'. Yet, Geddes concluded, 'there seem[ed] nothing for it but to try'.[18]

A second chance for civics

As the second of the two speaking slots Branford had secured for Geddes when the Sociological Society's programme of speakers was first being arranged in mid-1904, the meeting on the 23 January 1905 represented the final phase of Branford's original plan for the British sociology project. However, with Galton having already cast a vote of no confidence in British sociology, Geddes was in a position that neither he nor Branford would have envisaged a year earlier. After his disappointing paper of the previous July, Geddes was under a great deal of pressure to demonstrate that Branford's faith in him was not misplaced and, for this reason, Geddes tried to make his second presentation to the Sociological Society a vast improvement on the underwhelming first paper. Indeed, at three times the length of his first talk, Geddes' second presentation was intended to both correct the mistakes he had made earlier and establish he could provide a theoretically rigorous yet empirically grounded research programme for British sociologists.

With Charles Booth, the social statistician and famed author of the seventeen-volume *Life and Labour of the People of London*, in the chair, Geddes began his second presentation at the LSE in February 1905 by suggesting to his audience that the lukewarm response he had received from them six months earlier had mostly been a result of his having chosen a title, 'Civics as Applied Sociology', that bore little relation to what he had actually said. A better title would have been 'Civics as Concrete Sociology', he argued, because his aim had been 'to plead for the concrete survey and study of cities, their observation and interpretation on lines essentially similar to those of the natural sciences'.[19] Consequently, Geddes had changed the title of his second presentation to 'Civics: As Concrete and Applied Sociology, Part II' – an alteration that was supposed to clarify the link between his two papers and enable him to talk about what the application of his methods would mean for sociological practice.

Building on his earlier attempt to link biology and sociology, Geddes introduced civics as a social scientific method that was 'in line with the preliminary sciences, and with the general doctrine of evolution from simple to complex; and finally with the general inquiry into the influence of geographical conditions on social development'.[20] At the heart of this assertion was the comparison he drew for his audience between the biological dictum of 'environment – conditions – organism' and the 'place – work – folk' formula, which Geddes, inspired by the writings of the French social surveyor Frédéric Le Play, had made the foundation of civics.[21] What linked these two triads, Geddes told the Sociological Society, was their common explanation of individuals, whether they were people or simple biological organisms, as products of much greater wholes. In this sense, he told the Society, he wanted to

> press home the idea that just as the biologist must earn his generalisa-
> tions through direct and first-hand acquaintance with nature, so now
> must the sociologist work for his generalisations through a period of
> kindred observation and analysis, both geographic and historical; his
> "general laws" thus appearing anew as the abstract of regional facts,
> after due comparison of these as between region and region.[22]

In searching for these generalizations, Geddes went on to explain, the sociologist needed to pay attention to two different but interrelated and ontologically equivalent parts of society: its material components, such as individuals, institutions, industries, and events, which he called the 'town', and a corresponding 'thought-world', called the 'school', which was made up of the ideas and doctrines that provided the context for actions within the town.[23] Existing in a state that would not unfairly be described as a dynamic equilibrium, society developed, Geddes argued, through a process of constant adjustment between schools and towns, with change in one leading to a change in the other. Consequently, social development had to be understood as a complex, rather than monocausal or unidirectional, process, which made it imperative for sociology to become a science dealing with evolution throughout the social whole, from the organic form of individual people to the intellectual thought of the age.

Using the 'thinking machines' he had been developing since the early 1880s, Geddes outlined how the 'place – work – folk' formula could aid this style of sociological practice by providing a framework for social investigation. (See Figures 6.1–6.3.) Talking his audience through the various permutations and combinations of place, work, and folk, from the natural advantages of a particular area ('place-work') to the role of occupations in shaping the outlook of groups of people ('folk-work'), Geddes argued that an expanded and tabulated version of the 'place – work – folk' formula identified the various factors that sociologists needed to examine. However, he went on, this method had to be used in a historical, rather than a static, investigation because it was only by charting change over time in the various place, work, and folk categories that the sociologist would be able to understand how new social wholes emerged from older economic, geographic, biological, and social forms. 'While the complex social medium has thus been acquiring its characteristic form and composition', Geddes told the Sociological Society,

> a younger generation has been arising. In all ways and all senses, Heredity is commonly more marked than variation – especially when, as in most places at most times, such great racial, occupational, environmental transformations occur as those of modern cities. In other words, the young folk present not only individual continuity with their organic predecessors which is heredity proper, but with their social predecessors also...The younger generation, then, not only inherits an organic and a psychic diathesis; not only has transmitted to it the accumulations, instruments and land of its predecessors, but grows up in their [social] tradition also.[24]

Starting then once more with the simple biological formula:

ENVIRONMENT · · · · CONDITIONS · · · · ORGANISM

this has but to be applied and defined by the social geographer to become

REGION · · · · OCCUPATION · · · · FAMILY-type and Developments

which summarises precisely that doctrine of Montesquieu and his successors already insisted on. Again, in but slight variation from Le Play's simplest phrasing (*"Lieu, travail, famille"*) we have

PLACE · · · · WORK · · · · FOLK

PLACE FOLK ("Natives")	Work-FOLK ("Producers")	FOLK
PLACE-WORK	WORK	FOLK-WORK
PLACE	WORK-PLACE	FOLK-PLACE

Figure 6.1 Continued

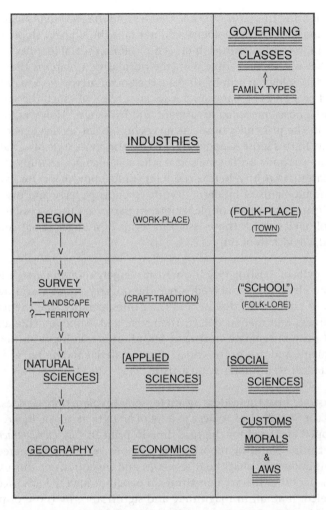

Figures 6.1 Geddes' translation of 'Environment – Conditions – Organism' into his "Place – Folk – Work" formula, which he had appropriated from the work of the French social surveyor Frédéric Le Play, and two of the 'thinking machines' that Geddes developed from that formula. These thinking machines were meant to illustrate the wide array of relationships between the ideas Geddes was discussing

Source: Geddes 1905: 71, 72, 77.

In adopting this basic outlook, Geddes told the Sociological Society, it was important that sociologists did not take the slippery slope towards a geographical determinist approach to human social identity – a position that was incompatible with the normative ambitions that underpinned sociology in Britain. Instead, the 'social survey' he had described should be seen as preparation 'for social service' in much the same way that 'diagnosis' preceded 'treatment and hygiene'.[25] However, in keeping with the principles of social surveying, social service needed to be holistic. In this sense, Geddes argued that the civics style of social amelioration was quite unlike the social reform campaigns that his audience were familiar with. Whereas the temperance movement, for example, saw the problems of modern city life, such as poverty and disease, as being caused by a few simple factors, in particular the excessive use of alcohol, civics recast those issues in terms of a larger social complex. Indeed, Geddes went on,

> the evils of existing city life are thus largely reinterpreted, and if so more efficiently combated; since the poverty, squalor and ugliness of our cities, their disease and their intemperance, their ignorance, dullness and mental defect, their vice and crime are thus capable not only of separate treatment but of an increasingly unified civic hygiene, and this in the widest sense, material and moral, economic and idealist, utilitarian and artistic.

To illustrate what these ideas meant for social explanation, Geddes introduced the Sociological Society to a set of analytic terms he had recently developed and used for the first time in print in *City Development*, his Dunfermline planning report of the previous year.[26] 'All are agreed', he told his audience, 'that the discoveries and inventions of this extraordinary period of history constitute an epoch of material advance only paralleled, if at all, in magnitude and significance by those of prehistory with its shadowy Promethean figures'. However, by utilizing the methods he had described, Geddes had reinterpreted those industrial advances as a process consisting of two different sociological complexes, rather than a single homogenous trend. The first, which he called the 'Paleotechnic' and saw as dominating the earliest stages of industrialization, was a 'comparatively crude and wasteful technic age, characterised by coal, steam, and cheap machine products, and a corresponding *quantitative* ideal of "progress of wealth and population"'. The second, which he called the 'Neotechnic' and identified with a number of recent trends, such as electricity, was an incipient stage of industrialization that was 'characterised', he explained, 'by the wider command, yet greater economy of natural energies … and by the increasing victory of an ideal of *qualitative* progress, expressed in terms of skill and art, of hygiene and of education, of social policy, etc'.[27]

In essence, Geddes argued, the whole of contemporary civilization, from philosophical and scientific ideas to technological artefacts, could be understood as belonging to one of these two descriptions. Whilst dividing the world up in this fashion served a purpose in providing a more nuanced understanding of social development, its real value lay, he claimed, in how it could be applied by sociologists to facilitate social progress. By virtue of the fact it was built around human interests and growth, the Neotechnic complex was clearly better than the Paleotechnic, which had caused many of the social problems the Sociological Society was interested in solving. However, although the Neotechnic phase grew directly from the Paleotechnic, it was by no means inevitable that Paleotechnics would be overcome. The triumph of Neotechnic methods, such as industries humans could control rather than submit to, was dependent on humans choosing them over their Paleotechnic ancestors. Thus, Geddes argued, the normative role of the sociologist was to help build a better society by identifying and recommending the uptake of Neotechnic ideas.

As he brought his presentation to a close by suggesting that these ideas could be disseminated through exhibitions of the kind he had developed at the Outlook Tower, Geddes and his supporters would have been anxious about what the audience had made of his undoubtedly ambitious talk. In opening the discussion, the meeting's chair, Charles Booth, complimented Geddes on how he 'seem[ed] to widen and deepen the point of view, and to widen and deepen one's own ideas'. Moreover, Booth reflected on how he found it 'interesting and instructive and helpful to follow [the] charming diagrams which spring evidently from the method [Geddes] uses in holding and forming his conceptions'.[28] No doubt to Geddes' great pleasure, the rest of the audience had picked up on and approved of one of his ideas in particular: the belief that sociology could and should empower humans to take control of the forces that had always shaped them. Indeed, as one contributor to the discussion put it, Geddes' paper gave them hope that 'the time [was] coming when we shall bring the force of our own characters to bear on our environment, and endeavour to break away from conditions which have made us the slaves of environment'.[29]

However, with the exception of the Rev. Dr. Aveling, who asked Geddes to be specific about 'the exact value to be given to the seemingly contradictory doctrines that the individual is the product of the city and also that the city is the product of the citizen', no one at Sociological Society engaged closely with the substantive content of Geddes' presentation.[30] More worryingly for Geddes, though, of the direct comments on his proposals that he did receive, very few offered unqualified support for the research programme he had put forward. In fact, as E. S. Weymouth's description of the civics social survey as a 'formidable task' demonstrated, most contributors to the discussion seemed overwhelmed by the sheer volume of material that Geddes had covered – something that highlighted how he had, if anything, overcompensated for his earlier failings at the Society.[31] In this sense, whatever his listeners had made of the methods and ideas Geddes had talked about, they were still to be convinced that civics was the best way forwards for sociology in Britain.

Aside from what anyone actually said at the meeting of January 1905, what testified most strongly to the fact that Geddes had fluffed his lines at the Sociological Society once again was the sudden deterioration of Branford's health. After almost two years of unpaid labour as the Sociological Society's secretary, Branford realized that British sociologists were not going to take up civics, just as they were not going to take up eugenics. The plan he had put into motion in mid-1903 had come to the end of the road and, as a consequence, Branford was not only short of ideas about what to do next but also short of the energy that the work would require. As he wrote in a letter of August 1905, Geddes, who was not usually known for his appreciation of his supporter's efforts, knew that he was at least partly 'responsible' for causing Branford's problems.[32] However, it is unlikely Branford welcomed Geddes' admission that he had played a role in destroying what remained of his chances to seize control of the British sociological agenda.

Reacting to this situation in April 1905, the Sociological Society's council voted, on the motion of L. T. Hobhouse, to relieve Branford 'of some of [his] heaviest work' by looking for someone who would do the job on a full-time and paid basis.[33] In a number of important ways, from its practical implications to its symbolic value, this decision marked the beginning of a new stage for the British sociology project. Although Branford had not stepped down completely, he had relinquished many of his duties, which signalled that the Society was going to be following a new path. The big question, though, was which path that would be. With Galton and Geddes' research programmes being judged unconvincing as visions of sociology's future, the pressure was on for the Sociological Society to find some kind of consensus to overcome the disagreements and disputes that had been a major feature of the its first two years. However, in its first response to this challenge, the Society was far from convincing.

Which way next? life after Geddes and Galton

After several months of recuperation and a business trip to Argentina in late 1905 and early 1906, Branford returned to the Sociological Society as its 'honorary secretary', a position that had been awarded to him in recognition of his status as one the most senior figures in the British sociology movement. In the meantime, though, the Society had recruited a new secretary, completed the scheduling of its meetings for the 1905–6 academic year, and issued the second volume of its transactions, the *Sociological Papers*. The person whom the council of the Society appointed as secretary in the autumn of 1905 was J. W. Slaughter, an American social psychologist from Clark University in Worcester, Massachusetts, who was chosen because he was 'specially trained in the social sciences, and familiar with contemporary work in the sociological field in Great Britain and abroad'.[34] However, to secure his services, the Society had been forced to improve the monetary value of the package on offer by arranging for Slaughter to lecture on social psychology at University College London. Whilst this lectureship was not formerly part of the Martin White Benefaction to promote sociology within the University of London, it did mean that the Sociological Society had yet another foot in the door of the academy. Indeed, it was a further boost to the news that the University of London had agreed to not only renew Hobhouse and A. C. Haddon's lecturing contracts for the following year but also establish two bursaries and a studentship for students studying on their courses.[35] The problem, though, was that none of these developments represented anything new; on the contrary, the driving force behind them was what little inertia remained from the plan that had guided the Society during the previous two years.

With both eugenics and civics having failed to capture the imagination of the Society's membership, it was clear the future that Branford had imagined for British sociology in mid-1903 was unlikely to be realized. Speaking at a meeting of the Society in May 1905, Hobhouse touched first on the issue that was dividing the British sociological project when he criticized 'the attempt to deal with the science of society as if it were a department of the science of biology'. For Hobhouse, an avowedly anti-biological thinker who had recently started work on *Morals in Evolution*, the final, and sociological, instalment of his trilogy of books on evolution, the 'object of the Sociological Society [was] to protest against that method of treatment, by insisting on the historical study of social phenomena. Man comes under the general scheme of biology', Hobhouse argued,

> in so far as he is a living animal and certain laws of biology are interesting to the student of affairs. But man is a peculiar animal. He is sometimes called a rational animal, though he might with equal point be called the irrational animal, since he is the only animal that voluntarily does what is opposed to his interests. To endeavour to predict the course of human evolution on the strength of analogies drawn from the animal and vegetable kingdom is to omit the dominant fact that distinguishes the human from the animal world.[36]

Yet rather than attempt to heal the divisions that had opened up on this question, the council of the Sociological Society managed to maintain them by pursuing a version of the strategy Branford had, to all intents and purposes, exhausted. In fact, despite Galton having signalled his disapproval of their activities, the Society even persevered with the effort to forge links with the eugenics project at UCL by encouraging their members to contribute to the 'Register of Able Families', on which Edgar Schuster, the Galton Fellow in National Eugenics, had started work.[37] Aside from the way it alienated people such as Hobhouse, the major problem with this decision not to tamper with Branford's template for attracting people to sociology, even though the ends that it had been designed to achieve were no longer obtainable, was that it created a genuine sense of intellectual instability. Indeed, after over two years of presentations, discussions, and lectures within the University of London, the Society was not only no closer to uncovering 'the conditions and forces which respectively hinder or help development, which make towards degeneration or towards progress' than it had been when it started but had also made little progress on the question of how to go about doing so.[38]

This ongoing uncertainty over the identity of sociology found its clearest expression in the programme of meetings the council of the Sociological Society had organized for the 1905–6 academic year. Although most of the presentations could be described as having something in common, specifically a concern with social problems and a belief that sociology could help solve them, there was a massive intellectual chasm between speakers on the question of how sociology should be practiced. On the one hand, there were papers, such as 'The

Biological Foundations of Sociology' by G. Archdall Reid, a surgeon and writer on heredity, and 'A Practicable Eugenic Suggestion' by William McDougall, a reader in mental philosophy at Oxford, which continued the discussion about eugenics and how to relate biology and sociology.[39] Indeed, after two years of trying, the Sociological Society was even able to secure a paper, entitled 'The Sociological Appeal to Biology', from Geddes' collaborator, J. Arthur Thomson.[40]

On the other hand, though, the Sociological Society's programme of events was packed with papers that had no concern with biology at all and which were instead more closely related to other social sciences, including economics and anthropology. For example, whilst in November 1905 the Society heard from the anthropologist A. E. Crawley on 'The Origin and Function of Religion', they also heard from the Fabian Beatrice Webb on 'Methods of Social Investigation' in January 1906, as well as the social reformer William Beveridge on 'The Problem of the Unemployed' at a conference on unemployment that the Society had organized at the LSE in April that year.[41] However, as the published records show, very few people moved between these very different types of meeting. For the most part, enthusiasts for the application of biology to social questions attended the eugenics sessions, whilst those who were not convinced by the need for biology in social explanation were present at the other meetings. Thus, far from bringing social investigators together, the Society's discussions were continuing to divide them.

Of all the meetings that took place at the Sociological Society during the third year of its existence, there were two that summed up the extent to which the hope that had driven the organization in its early days had faded. The first was a presentation from Geddes, who spoke in March 1906 under the heading 'A Suggested Plan for a Civic Museum (or Civic Exhibition) and its Associated Studies'. Elaborating on the claim with which he had ended his paper of January the previous year, that exhibitions could link the intellectual debates of sociology and its practical aims, Geddes talked about his belief that 'Civic Museums', such as the Outlook Tower, could help people understand sociological generalities, such as his Paleotechnic and Neotechnic ages.[42] However, this third presentation, which was given only two days after Thomson's, was badly received, even more so than Geddes' previous two. According to the brief record of the discussion, which, perhaps tellingly, failed to note who took part, Geddes' audience was confused about why the kinds of exhibitions he had described were necessary at all. Moreover, they also found his attempt to relate theoretical ideas and practical ends ambiguous. As Lady Welby put it in a letter to Branford two months later, Geddes' lecture confirmed that his efforts in London were so 'futile' that the only rational course of action was for him to give up.[43]

The second session that evidenced the problems at the Sociological Society during the 1905–6 academic year was the occasion of the first presentation by the writer H. G. Wells, a one-time member of the editorial committee of the *Sociological Papers*. Speaking under the heading 'The So-Called Science of Sociology', something that would never have been entertained by the Society when it was first founded, Wells surveyed the history of both the general idea of sociology and the Sociological Society itself. In so doing, he argued that, contrary to popular opinion, 'no sociology of universal compulsion, of anything approaching the general validity of the physical sciences, is ever to be hoped for'.[44] 'A review of the first volume of the [*Sociological Papers*] brings home the...image', Wells told the Society, 'of exploratory operations in "taking a line"'. However, he went on, the list of contributors stirred memories of 'works that impress one as large-scale sketches of a proposed science rather than concrete beginnings and achievements'. The problem, though, was that 'the search for an arrangement, a "method", continues as though it were not'.[45] In this sense, Wells argued, it could only be concluded that 'sociology must be neither art simply, nor science in the narrow meaning of the word at all but knowledge rendered imaginatively...that is to say...literature'.[46]

With ideas such as these being aired, the Sociological Society in 1906 was a very different place to what it had been in 1903. Although the teaching of sociology within the University of London, which the Society had helped establish, was still expanding, as James Martin White's decision of mid-1906 to renew the Finnish anthropologist Edward Westermarck's post as lecturer for a further five years demonstrated, there was no clear sense any more of where the whole enterprise was heading.[47] It was clear at the end of the 1905–6 academic year that the British sociological project was divided and that there was no obvious way of bringing its various factions back together. Indeed, in the absence of the kind of unity that had characterized the Sociological Society's early years, it seemed doubtful whether there was either the enthusiasm or the scope for developing sociology in a way that would achieve the aims the Society had set out in late 1903. Something needed to be done if the British sociological project was not to come to a grinding halt. Recognizing this fact, Branford stepped forward with a new idea for unifying the enterprise he had had helped to create.

L. T. Hobhouse: Martin White professor of sociology

On the 15 January 1907, Branford wrote a letter marked 'private & confidential' to Lady Welby. The reason for his writing, Branford told Lady Welby, was the news that Hobhouse had become 'practically free' of his duties as political editor of the daily newspaper, the *Tribune*. 'Now' this had happened, Branford explained to Lady Welby, he was of the opinion that 'something ought to be done to secure [Hobhouse] for Sociology'. After four years of meetings and debates at the Sociological Society, there was a widespread sense the British sociology project was on the verge of failure. However, Branford believed that Hobhouse was the 'one personality round which the whole movement' might regain the cohesion and momentum it had once possessed. As a consequence, Branford had conceived of what he called a 'general scheme' to bring Hobhouse and sociology together. Moreover, despite the fact he had yet to approach Hobhouse about this plan, Branford had begun to put it into motion.[48]

Branford had a number of reasons for homing in on Hobhouse as the person who could solve the problems of the British sociology project. Whilst he had not given a presentation to the Sociological Society during its formative discussions, as both Geddes and Galton had done, Hobhouse had been involved with the Society since its inception, serving on both its council and the editorial committee of the *Sociological Papers*. Moreover, Hobhouse had taken part in some of the Society's most important and controversial debates. For example, when Galton had first presented to the Society on his programme for eugenic research in May 1904, Hobhouse had made one of the most eloquent and principled of the session's speeches about the importance of maintaining a sceptical attitude towards the claims of biology on the social sciences.[49] Indeed, in addition to elaborating on this idea at subsequent meetings of the Sociological Society, Hobhouse had recently published *Morals in Evolution*, the third and final instalment of his series on evolution, discussed in chapter four, in which he had argued for an evolutionary account of society that was independent of the evolutionary views of biologists. Given the problems he had encountered in attempting to sell the ideas of Galton and Geddes to the Sociological Society, it was no surprise Branford thought British sociology might be unified, as he told Lady Welby, through a 'Sociological Faculty' in which 'a Chair or Lectureship of General Sociology for Hobhouse' was the focus.

Yet, as Branford's initial suggestion for an institution to house this faculty demonstrated, his thought was still in many ways rooted in the ideas that had consumed him during his failed attempts to pro-mote eugenics and civics during the previous four years. 'University College.... is the one University Institution in London', Branford told Welby, 'of the first culture rank'. It therefore seemed 'appropriate', he argued, 'to get the beginnings of a Sociological Faculty established there rather than elsewhere'. After all, Branford went on, 'the beginnings of such a Faculty exist in (a) the Research Fellowship for Eugenics, (b) Slaughter's Lectureship in Social Psychology, and (c) [the psychologist, philosopher, and member of the Sociological Society's provisional com-mittee, James] Sully's lectureship in Aesthetics'. There were, however, two major obstacles for Branford to overcome if these existing positions were to be turned into a new faculty. The first was to secure the approval of both Hobhouse, who would not welcome an association with eugeni-cists, and the University of London, who would have the final say over institutional arrangements. The second problem was money, around £15,000 of which Branford estimated the new faculty would require.[50]

As he told Lady Welby in a letter of the 22 January 1907, Branford decided in the first instance to attempt to establish what the University of London thought of his idea. He did this by discussing the 'project with [his] old friend Joseph Hartog, who as Academic Registrar of London University [had] administrative control of the Martin White and the Galton benefactions'. To Branford's delight, Hartog, who had been a colleague of Hobhouse's at the *Manchester Guardian* during the 1890s, considered the suggestion of setting up a sociological faculty to be a good one.[51] However, as Branford explained to Lady Welby, Hartog had also counselled him to make two alterations to his pro-posal. Firstly, Hartog had told Branford that the plan should 'hinge on getting a chair (not a lectureship) of Sociology' within the University of London. Secondly, and most importantly, Hartog had strongly advised Branford to reconsider his choice of home institution for the faculty. Indeed, Hartog was of the opinion that a chair of sociology should not be established 'in connection with University College'.[52]

Following these recommendations, Branford removed all reference to UCL from his proposals and considered what further adjustments this change would require. Moreover, he began to contact influen-tial supporters of the British sociology project who he believed could help him promote the revised plan. As he explained in a letter of early February to James Bryce, the Liberal politician and former president of the Sociological Society, Branford had decided that his proposed socio-logical faculty should feature 'Dr. Westermarck, Professor Haddon, Mr. L. T. Hobhouse, and Dr. J. W. Slaughter' – the 'four sociologists' who had been 'temporarily' attached to the University of London via 'the three years experiment (under the Martin White Benefaction)'. On Branford's account, Westermarck, Haddon, Hobhouse, and Slaughter represented 'the best tendencies in contemporary sociology' and thus, Branford told Bryce, they would form 'a University Department of Sociology, such as few, if any, other Universities have'. 'In a lectureship of Ethnological Jurisprudence', Branford explained, Westermarck

> would bring to the department some of the best German tradition – the approach to Sociology from the side of Law and Comparative Ethics; Hobhouse in a lectureship of General Sociology, would continue the older English tradition (from Hobbes to Mill) of an approach from a general social and political survey; Haddon in a lectureship of Anthropology would represent the newer English approach from the side of Natural Science; while Slaughter in a lectureship of Social Psychology, would ably and fully represent that approach from Psychology to which the Americans have perhaps contributed most.[53]

Suitably persuaded by Branford's argument, Bryce agreed to recommend to the University of London that they 'establish a department of Sociology with Lectureships for Westermarck, Hobhouse, Haddon, and Slaughter'. 'Of course', Branford observed to Lady Welby, there was no reason for him to think 'anything [would] come of Bryce's' advocacy. However, Branford concluded, it did seem to be a moment 'at which we must halt for a time, and await the effect'.[54]

Whilst Bryce appealed to the University of London on his behalf, Branford set about trying to solve the second of the two problems that had confronted his original proposal: money. As it was on every other occasion he needed funds for the British sociology project, Branford's first port of call on this matter was James Martin White, who Branford hoped could be convinced to pay for a sociological faculty in much the same way he had paid for the Sociological Society. Since he had first conceived of the faculty idea, Branford had estimated he would need to find upwards of £15,000, approximately £1,150,000 in early twenty-first century money, to make it a reality.[55] Yet, despite, or perhaps because of, having already spent several thousand pounds on the British sociology project, White could not be convinced to hand over the £15,000 Branford wanted. White did agree, though, that much of what Branford had proposed was a good idea and agreed to provide the not insubstantial sum of £10,000 for the purpose.

With these funds secured, Branford decided in late April that the time was right to make an official approach to the University of London. The strategy to do this he settled on was to have the Martin White Benefaction Committee prepare a report, which they approved and forwarded to the University of London on the 5 June. Given the amount of funding Branford had secured was somewhat less than he had sought, the proposal the University of London received was less ambitious in scope than the 'scheme' Branford had envisaged the previous January. Gone were both the notion of setting up an actual department of sociology and, with it, the suggestion a job should be found for Slaughter, the most recent arrival on the British sociology scene.[56] Yet, as the Committee's report and a letter from White that accompanied them made clear, the three-part plan for spending White's £10,000 kept to the spirit, if not the letter, of Branford's original ideas.

The first, and most significant, aspect of the Committee's package of proposals was for the University of London to found 'a Chair of Sociology', which, White explained in his covering letter, 'should be offered, in the first instance, to Mr. L. T. Hobhouse'.[57] The second proposal was for what remained of the £10,000 to be used to make Westermarck a 'University Professor of Sociology' for five years and Haddon a 'University Lecturer on Ethnology'.[58] Finally, and no less importantly, the University was being asked by the Committee to house Hobhouse and other teachers of sociology at the London School of Economics.[59] As the various enclosures that were attached to these requests as they made their way to the University's Academic Council and Senate explained, the permission to proceed with this plan had been obtained from all the relevant parties, including the LSE. It was therefore up to University of London to either approve or reject what had been put before them.

Admitting they were unable to put together a group with more expertise on sociology than the Committee from which these proposals came, the University of London resolved at a meeting of its Senate on the 24 July that 'the donations offered by Mr Martin White for the endowment of Chairs in Sociology be accepted under the conditions' that had been stated.[60] Indeed, in further recognition of the fact they would be unable to find anyone better qualified to assess the merits of candidates to fill such posts, the University even waved their usual administrative procedures, which ensured they could monitor and exercise control over academic appointments, so that they could offer Hobhouse and Westermarck the jobs straight away. Thus, a little over six months after it had first been mooted, Branford's scheme to save the British sociology project had become a reality.

After four long and frequently controversial years of debate, those involved in British sociology had finally achieved what they had set out to do when they first explored the idea of forming the Sociological Society in mid-1903. In the chair and professorship at the LSE, formal recognition for sociology in the curriculum of the University of London, the Sociological Society itself, and in the *Sociological Papers*, which would soon be renamed the *Sociological Review*, Branford and others had helped establish a basic disciplinary framework for sociology in the UK. However, in selecting Hobhouse as the first occupant of the Martin White chair, British sociology had arrived at an intellectual identity that was somewhat different to the one that had seemed most likely when discussions about the content, scope, and aims of the subject had first taken place at the Sociological Society. Through its critical reception of Galton's eugenic programme, scepticism towards Geddes, and decision to install Hobhouse as its *de facto* leader, the British sociology community had rejected the idea that there should be substantive links between sociology and biology. For this reason, Hobhouse was in an unrivalled position with respect to shaping the academic agenda for sociology in the UK and, as we will now see, he gave a first glimpse in late 1907 of the hugely consequential ways he intended to do so.

Conclusion

On the 17 December 1907 L. T. Hobhouse appeared before an audience including Sir Arthur Rücker and Sir William Collins, the principal and vice-chancellor of the University of London, at the London School of Economics to deliver his inaugural address as Martin White professor of sociology. As an event that marked both the end of one chapter in the history of British sociology and the beginning of another, Hobhouse's address, which was entitled 'The Roots of Modern Sociology', had one major aim: to get a 'clear conception of what [sociology] is, and what lies before us in the study of it by tracing its roots in the history of thought'.[1] However, despite striking a conciliatory tone that was worthy of someone who was supposed to unify British sociology, Hobhouse also wanted to declare an end to the debates about the place of biology in social science, which had threatened to consume the Sociological Society during the previous four years. In so doing, he not only gave a succinct summary of this book's arguments about the origins and content of the founding British debates about sociology but also highlighted why he, rather than Patrick Geddes or a disciple of Francis Galton, had secured the UK's first chair of sociology.

Hobhouse told his audience at the LSE that, though there were arguably more, there were four roots that were most significant when it came to understanding the state of and hopes for sociology in early twentieth-century Britain: political philosophy; the philosophy of history; the natural sciences; and the tradition of political economy associated with Adam Smith. Political philosophy was significant, Hobhouse argued, because its practitioners, from Thomas Hobbes to T. H. Green, had analysed 'the actual conditions under which we live' so they could work 'out an ideal by which or for which society should live'.[2] For Hobhouse, who had forged his philosophical first principles and political sensibilities

170

in the midst of the idealism of Green and others, this aim made political philosophers to earlier generations what sociologists were aspiring to be to the present and future: guides for social action. However, Hobhouse went on, the form of sociological, as opposed to political, explanation had been shaped by other considerations, which would have been familiar to anyone who had followed the debate about sociology since the late 1870s controversy over the place of Section F at the British Association for the Advancement of Science. Indeed, sociologists were the heirs, he explained, of Adam Smith's aspirations for a general science of society, which had, of course, been thwarted by Smith's 'spiritual descendents' who had pursued 'economics in isolation from the study of other social phenomena'.[3] Sociologists were going finally to fulfil Smith's original plans for political economy, Hobhouse argued, but they were going to do so having learned the lessons of past debates and thus make sociology a historical, rather than deductive or analytic, discipline. In this sense, Hobhouse reasoned, the identity of sociology was related in part to the philosophy of history; that is, to the idea that social change could only be explained with reference to mechanisms or forces that acted over time.

Yet despite these connections with other fields, Hobhouse told the LSE that sociology also had an important but more difficult relationship with the natural sciences. 'Whatever science is most flourishing and progressive acquires a preponderating influence', he explained, 'over the general thought of the time'.[4] For this reason, whilst social scientists of the past had tried to model their work on physics, a great many sociologists had recently fallen under the spell of biology. In fact, 'social questions [had come] to be referred to the biological principles which underlay the life of society', so that

> it was held that the struggle for existence, natural selection, and the survival of the fittest, were the key to all possible progress on this earth. In any question of statesmanship ... the real question to decide was whether the measure proposed would or would not tend to arrest the struggle for existence, to preserve the unfit, and so blunt the edge, as it were, of the keenest weapon which nature had provided for the advancement of the species.[5]

However, in the future, Hobhouse declared, sociologists were not to feel as though they had to defer judgment on social questions to the biological sciences. Biology and sociology 'are quite distinct', he argued, and the identity of sociology was dependent on sociologists maintaining this separation. In fact, he went on, the future of British sociology was one in which sociology was going to stand in relation to biology as biology did 'to physics and chemistry'.[6]

Just a few months later, Hobhouse was given the opportunity to repeat this message in his first editorial for the *Sociological Review* – a new quarterly journal, which had been established by the Sociological Society as a replacement for the *Sociological Papers*, which had been published between 1904 and 1906. Building on the tone and content of his inaugural lecture, Hobhouse explained to readers of the *Sociological Review* that he had no intention of being dictatorial when it came to deciding what did and did not have a place in the only British journal dedicated to sociology. However, it needed to be clear from the outset, he asserted, that the *Sociological Review* was not going to fall into the trap that had nearly brought the Sociological Society to a grinding halt during the previous four years. For this reason, Hobhouse wrote, it was the duty of the *Sociological Review* to 'afford [sociologists] collectively the opportunity of defending the study of society from aggressions which would destroy its character as a distinct science or from usurpations which would merge it in the work of one of its own departments'.[7] Thus, whilst biological ideas had played an important role in getting British sociology to the point it had reached in late 1907 and early 1908, Hobhouse's intentions for the field were clear: it was to be an autonomous science that would not submit to the authority of any other field, least of all biology, when it came to making sense of social phenomena.

What difference did Hobhouse make to British sociology?

Although there are a number of points at which one might chose to end an account of the early history of British sociology, Hobhouse's inaugural lecture at the LSE and his editorial for the *Sociological Review* seem the most appropriate. This book has been concerned with explaining why it was Hobhouse, rather than Patrick Geddes or a candidate to Francis Galton's liking, who was given the opportunity to shape British sociology's identity and sense of purpose in the early twentieth century. As the previous six chapters have shown, the reason was Hobhouse's conviction that sociology should be separated from biology – a fact that has been overlooked in the historiography of British sociology, which focuses mainly on what came next: the alleged 'failure' of the discipline. In this sense, finishing with Hobhouse's first acts as the *de facto* head of British sociology is fitting because it enables us to take stock of events without feeling they need to be evaluated in terms of what did, or rather did not, happen next. More specifically, Hobhouse's statements of his intentions for British sociology provide an opportunity to evaluate the significance of his appointment by asking what would have been different if it had been Geddes or a disciple of Galton's who had been able to set the discipline's agenda in late 1907 and early 1908.

Although historians have frequently seen them as the kind of question that belong to those who would reduce history to radical contingency, counterfactual propositions such as these, which are taken far more seriously in the natural and social sciences than they are in the humanities, have a real value in testing our assumptions about the significance of particular events and the intentions of those who were involved.[8] Indeed, as scholars including Niall Ferguson, Gregory Radick, John Henry, Peter Bowler, Steven French, and Steve Fuller have all argued, there are a number of reasons why well-formulated counterfactual questions should have an important place in historical investigation. On the one hand, as Ferguson has suggested, counterfactuals convey a crucial point about history: that the actors involved were confronted with a range of possibilities they evaluated before settling on the course of action that was actually taken.[9] On the other, as Radick, Henry, Bowler, French, and Fuller all highlighted in a recent *Isis* focus section, considering what might have been is a crucial part of understanding what is significant and, in some cases, inevitable about the type of science we now have.[10] In this sense, when counterfactual reasoning is applied appropriately, it uncovers what historical actors understood to be at stake when they fought for a particular idea or theory and why a decision, such as the one to appoint Hobhouse at the LSE, was and continues to be important.

With respect to the history that has been explored here, one could conclude, as the sociologist Edward Shils did in the remarks that opened this book, that counterfactual considerations are largely irrelevant when it comes to assessing the development of sociology in the UK because the sociological imagination, which scholars frequently have seen as a prerequisite for the discipline's expansion, was never present in early twentieth-century Britain. However, what this book has shown is that, far from being absent from late nineteenth- and early twentieth-century Britain, sociology was the subject of an ambitious and theoretically sophisticated discussion. Moreover, this book has also demonstrated that choosing Hobhouse for the positions at the LSE and the *Sociological Review* was a commitment to a particular understanding of what would and would not have a place in British sociology from that point onwards. In this sense, whilst we cannot suggest whether

Galtonian or Geddesian sociology would have been any more successful than Hobhouse's programme, we can say that they are evidence that
two serious alternatives were on offer for sociology in early twentieth-
century Britain and that the intellectual identity of the discipline would
have been significantly different if either of them had been pursued.

The most concrete evidence for how different things might have
been for British sociology is to be found with respect to Galton's eugenics research programme. As we saw in chapter five, Galton established
a Fellowship in National Eugenics and a Eugenics Record Office at
University College London after deciding the Sociological Society was
insufficiently enthusiastic about his ideas. However, through the efforts
of his supporters, as well as Galton's bequests to UCL in 1911, a stronger
platform was built for eugenics within the University of London during
the early twentieth century. Whilst the Eugenic Record Office became
the Francis Galton Laboratory for the Study of National Eugenics in
1907, a chair of eugenics was created at UCL in 1911, which was occupied first by Karl Pearson, who also took charge of the laboratory, and
subsequently held by statisticians and eugenicists including R. A. Fisher
and Lionel Penrose. Furthermore, the Sociological Society's decision to
eschew eugenics in favour of Hobhouse's programme also proved to be
a catalyst for the popular side of the eugenics movement as it was a
group of disillusioned Sociological Society members who formed the
Eugenics Education Society in 1907. Later known simply as the Eugenics
Society, the organization was the popular front for eugenics in early and
mid-twentieth-century Britain and was known for its championing of
contraception and voluntary sterilization as means of improving the
physical condition and intellectual capacities of the UK's population.[11]
Given that Galton provided the finance and the agenda for the developments within academia and, despite his initial scepticism of their activities, served as honorary president of the Eugenics Education Society, it
is reasonable to conclude that what came to be known as 'eugenics' in
early and mid-twentieth-century Britain would also have been known
as 'sociology', had Galton persevered with the subject. Indeed, far from
urging sociologists to keep their distance from biologists, as Hobhouse
did, a Galtonian would have urged them to bring their work closer
together in late 1907 and early 1908.

However, as chapters three, five and six indicated, the question of
what might have been different about British sociology under Geddes is
fiercely debated in a way that it simply is not when it comes to Galton.
What makes assessing Geddes' intentions for British sociology difficult
is that he, unlike Galton, left no substantive institutional legacy and,

despite frequently promising to do so, never wrote a monograph on sociology, which meant he died without providing a definitive statement of how he believed the science should be practiced. Contrary to the strategy pursued by Galton, though, Geddes and his supporters, including Victor Branford, who continued to be Geddes' closest friend and collaborator, despite what some might interpret as a betrayal in 1907, remained a part of the Sociological Society and tried to alter the course of British sociology during the 1910s and 1920s. Yet for reasons that have still to be documented adequately, Hobhouse froze Geddes out of sociology at the LSE, which was the only university in the UK to have a chair of sociology before the expansion of the field after World War Two. In fact, even when Hobhouse resigned from the *Sociological Review* in 1910 after a dispute over his editorial policies and Branford took over, Geddes' cause was not helped because the journal was simply detached from developments at the LSE. As a consequence, Geddes was forced to develop his sociology during the first three decades of the twentieth century as he had done during the last two of the nineteenth: away from British universities and through organizations, such as the seldom-studied Le Play House, that were consciously geared towards his ideas.[12]

Despite these troubles, Geddes and his sociology did continue to attract followers during the later stages of his career. By far and away the most prominent but also independently minded of these supporters was the American writer Lewis Mumford, who, despite a difficult relationship with an increasingly cantankerous Geddes, promoted Geddesian regionalism in the USA through a range of activities, including his journalism for the *New Yorker* magazine and his involvement with the Regional Planning Association of America.[13] Indeed, as Richard C. Gunn and I have argued elsewhere, Mumford's writings, from his 'Renewal of Life' series, which included the explicitly Geddesian *Technics and Civilization* and was published between 1934 and 1951, to his famed 'Myth of the Machine' series of the late 1960s and early 1970s, exhibit clear evidence of Mumford having appropriated the principles of sociology Geddes had set out in papers such as the ones he gave to the Sociological Society in 1904 and 1905.[14] In this sense, if we wish to understand something of what British sociology might have become had Geddes been awarded the Martin White chair in 1907, we could point to Mumford's sophisticated critique of what he called 'megatechnical' society, which was embraced by political radicals in Cold War America, and his writings on cities, such as *The City in History*, which won the American National Book Award for non-fiction in 1962.

Yet the suggestion that Mumford, a public intellectual with no training in or apparent connection to biology, can be taken as an indication
of where Geddesian sociology might have gone would seem to under-
mine the claim, made here and by Steve Fuller, that British sociology
under Geddes would have been built on biologically engaged foundations.[15] There are, though, a number of issues that force us to take seriously the connections between sociology and biology under Geddes. As
this book has shown, the programme Geddes took to the Sociological
Society in the early twentieth century was firmly grounded in biology,
both methodologically and in the way that he encouraged sociologists
to look closely at the biological roots of social identity. Indeed, as chapter three argued, an important part of understanding Geddes is grasping that his move towards the social sciences after the early 1880s did
not mean he left the natural sciences behind. Even after 1908, when
he was increasingly known for his work on town planning, Geddes
continued to publish on the life sciences and his last book, which was
co-authored with J. Arthur Thomson and published in 1931, was a massive two-volume treatise entitled *Life: Outlines of General Biology*.[16] As a
trained biologist, Geddes took biology seriously and, as Mumford put
it in the mid-twentieth century, Geddes conducted his work on civics
not as 'a bold innovator in urban planning, but as an ecologist, the
patient investigator of historic filiations and dynamic biological and
social interrelationships'.[17]

Mumford's unqualified statement of Geddes' biosocial credentials
does bring us to a point, though, that is crucial to any claim about
the connections between biology and society in civics: the fact that
the biology in which Geddes sought to ground social explanation was
the kind that he himself had practiced in the late nineteenth century.
As was observed in chapter two, the identity of biology since the early
1900s has come to look a great deal more like the science practiced
by Galton's followers than it has the Spencerian science that excited
Geddes. In this sense, Geddes' biology and his understanding of where
the line was drawn between the natural and social worlds can be confusing to us now in the early twenty-first century, especially to those
who are unfamiliar with the intricacies of late Victorian biological discussion. However, just because Geddes' work bears little resemblance to
early twenty-first-century biology does not mean civics was not biologically engaged. On the contrary, Geddes' sociology was underpinned by
a consciously non-Galtonian form of biology and, thus, on his watch,
sociology would likely have involved a constant engagement with the
kind of biology described in chapter three – a fact that may one day

be demonstrated by a close study of the unpublished manuscript on biology that occupied Mumford during the final years of his life.[18] For these reasons, whilst it is unlikely that Geddes and his supporters would have identified themselves with the Nazi party platform, as Fuller has claimed, they would certainly have connected biology and sociology in ways scholars of the post-World War Two era have frequently found difficult to come to terms with in seemingly progressive thinkers.[19]

Whilst Galton, Geddes, and Hobhouse had a number of things in common, such as their normative ambitions for social science and their emphasis on the potential for human agency to alter the course of evolution, it was these fundamental differences in their understanding of how to relate biology and social explanation that would have meant a significantly different identity for British sociology, had it not been Hobhouse who was awarded the Martin White chair of sociology at the LSE and the editorship of the *Sociological Review* in 1907 and 1908. Only Hobhouse wanted British sociology to be free from biological ideas and, in this sense, he made a very real difference to the direction the discipline took from the moment that he was appointed. Indeed, it is only by recognizing the importance of this late nineteenth- and early twentieth-century debate about the place of biological ideas in sociology that we can begin to meet the interpretative challenge British sociology has always seemed to pose for historians of the social sciences.

Reappraising the foundations of British sociology:

Why Galton, Geddes, and Hobhouse still matter

As the introduction made clear, a major aim of this book has been to ask whether sociology really was as intellectually weak in early twentieth-century Britain as received views have always suggested it was. Whilst we noted how critics such as Talcott Parsons and Perry Anderson have argued the UK failed to participate in the late nineteenth- and early twentieth-century discussions that shaped the discipline, we also saw how revisionist scholars have challenged the implications of those views by exploring what the American historian Harry Elmer Barnes once called the 'large amount of work on the border line of sociology' in Britain.[20] By building on those revisionists' efforts, this book has shown, contrary to the beliefs of even the most sympathetic of scholars, that there was a theoretically sophisticated debate about sociology in the UK. However, in so doing, this book has also shown that what concerned the participants in those discussions were issues that seldom feature in the historiography of sociology as a whole. In this sense, it has been argued that although the dominant features of the British debate about the identity of sociology were not typical of what one might find in contemporary France, Germany, or the USA, we should not assume, as received opinion often has, that the UK was somehow intellectually inadequate. Yet even with these crucial aspects of the British debate about sociology revealed, the question that might still be asked is why we should want to rethink our attitude towards the subject at all. At the risk of sounding jingoistic, there are a number of reasons for reappraising the history of British sociology that has been documented here and these reasons make that history of particular relevance to discussions about the uses and practice of writing histories of science.

Aside from the fact that the history of any discipline is a crucial part of understanding it, the debates that took place at the Sociological Society are important because they speak to the present in two different but interrelated ways. On the one hand, disputes such as those between Steve Fuller and Maggie Studholme, John Scott, and Christopher Husbands over the correct way to interpret the Geddesian tradition of sociology demonstrate that there are scholars who still care enough about figures such as Geddes and Branford to argue about their work. Indeed, those scholars not only care about recovering ideas that have been lost to British sociology for the best part of a century but also believe that doing so can contribute to debates that are taking place right now in sociology; a conviction that helps throw light on the second sense in which the events that have been studied here are important. Since the mid-1970s, when E. O. Wilson's book *Sociobiology: The New Synthesis* made the kinds of connections between biology and society that respectable scientists had avoided since the end of World War Two, the social sciences have frequently been immersed in controversies, such as those about evolutionary psychology, which have been focused on the question of where to draw the line between biological knowledge and social explanation.[21] However, to understand why those debates have taken place at all and what has been controversial about the work of Wilson and others, we have to understand the origins of the relationships that sociobiology, evolutionary psychology, and other biologically engaged enterprises have challenged.

This connection between the debates that have been examined in this book and the discussions that are taking place right now in British sociology can be detected in the writings of the sociologist W. G. Runciman, a former president of the British Academy who has become widely known in recent years for his calls for sociologists to embrace a

Neo-Darwinian paradigm.[22] According to Runciman, sociologists of the twenty-first century need to face up to the fact that the classical generation, to whom they have always appealed, can no longer furnish sociology with the concepts and methods that are required to understand modern societies. Instead, Runciman argues that in light of the success of the modern evolutionary synthesis in biology, sociologists should accept that the thinker with the most to say about human behaviour is not Emile Durkheim or Max Weber but Charles Darwin. Although Runciman has constantly stressed that his claims need not lead to the crude reductionism of Darwinian social science of the past, his message for twenty-first century sociology has been clear: it is for sociologists to 'forget their founders' and instead embrace Darwin as the hero of modern social explanation.[23]

However, it seems unlikely that a conversation about the need for a Neo-Darwinian paradigm would be taking place in British sociology, or at least not on the same terms, if Galton or Geddes, rather than Hobhouse, had taken control of the discipline in 1907. Given that Hobhouse was the only British professor of sociology during his lifetime and was succeeded at the LSE in 1929 by Morris Ginsberg, who demonstrated his Hobhousean credentials by helping draft the 1950 UNESCO statement on race, which famously declared that race is a social not biological category, the division between natural and social worlds was rigidly maintained in British sociology throughout the first half of the twentieth century.[24] By the 1950s, when sociology began its first major expansion in the UK since Hobhouse's appointment, the tide had turned against biosocial science not just because of the Holocaust but also because of the programmes of forced sterilizations that had taken place throughout Europe and the USA during the 1930s and 1940s. It is therefore no surprise that the discussion of biology has been absent from mainstream British sociology until the late twentieth century.[25] Indeed, given Hobhouse's words of warning to his fellow sociologists in 1907, it should be no surprise that one of the major reasons given for accepting a Neo-Darwinian paradigm in sociology is the enormous success of modern biology, which, it is suggested by Runciman, amongst others, makes the current discussion very different to those that have taken place before.[26]

Yet as Runciman's evoking of sociology's founders reminds us, sociologists tend to think of themselves as being in touch with the origins of their discipline and, as a consequence, continue to engage with and teach the work of thinkers such as Durkheim and Weber. In this sense, there is every reason to argue that the British sociologists of the late nineteenth and early twentieth centuries can be recovered to help sociologists with the challenges they face in the early twenty-first century. Although the debates at the Sociological Society may have been about evolution, rather than genetics, their intellectual centre of gravity was exactly the same as the discussions that have resulted from the success of biology in the late twentieth century: how should sociology, as a general science of society, relate to biology, as a general science of life? In Galton, Geddes, and Hobhouse we have three different ways of negotiating the complex issues that come with answering that question. But, given that sociologists are used to looking backwards so that they can go forwards, it is quite possible that they could find in late nineteenth- and early twentieth-century British sociology the resources necessary to tackle the problems biology seems to pose for them in the present. Indeed, as Steve Fuller's call for sociologists to reconnect with Hobhouse's arguments for the separation of the biological and social sciences demonstrate, this process is one that has already started.[27]

History of science and the history of British sociology:

Integrating ideas and practice

Yet whilst it is important for us to consider why the events explored here are more than a parochial concern, this book has been a work of history rather than sociology. As the preface and introduction explained, this book's effort to recover the ideas and arguments at the heart of the founding British debates about sociology has been done with a particular set of historiographic trends, specifically the turn towards material practice, in mind. In closing, it is therefore appropriate for us to shift our attention towards some of the broader points this book has endeavoured to make about the writing of the history of science and to state more clearly the case for the alternative, intellectually orientated history that has been outlined here.

As the introduction highlighted, history of science has undergone an important set of changes over the course of the past two decades. During that time, the subject matter of history of science has been redefined as something called 'practice': an idea that many historians of science see as moving the field on from older and problematic philosophical or history of ideas orientated histories. In the most general sense, this book has embraced that conception of science-as-practice. Exploring how Galton, Geddes, and Hobhouse put together their programmes for British sociology, we saw how each of them was engaged in a project that involved action, from designing experiments to practical interventions in the places where they lived, as well as thought. Consequently, the British debate about sociology has been framed as a competition in which the prize was the opportunity to shape sociological practice from the ground up. However, in adopting that framing, this book has been clear that Hobhousean, Geddesian, and Galtonian practices were shaped by an engagement with particular sets of ideas. Rather than seeing Galton, Geddes, and Hobhouse's intellectual activity as being in some way divorced from their practical pursuits, this book has underscored how a crucial part of understanding what motivated them, as well as a wide range of others who were involved in the debates that culminated at the Sociological Society, was a belief in the power of the world of ideas to inspire and guide the world of action.

In this respect, as this emphasis on the connection between thought and action should suggest, this book has exercised scepticism towards some of the implications that the conception of science-as-practice has had for the general historiography of science. Specifically, by eschewing the emphasis on material culture that pervades contemporary history of science scholarship, this book has argued that if we want an analytically rich understanding of scientific practice then we need to take ideas, not just objects, seriously. At a time when many historians of science and their students have grown accustomed to deriding intellectual history as something that consists of 'elite' ideas that have little bearing on how science is actually done, this point about the way that people's actions are motivated by ideas is one that is worth repeating. Whilst subjects such as instrument makers and the communication of science through the periodical press have complicated our view of science in interesting ways, it is far from clear, as Christopher Hamlin has argued, that the recent enthusiasm for materiality in history of science scholarship compels us to discard long-standing notions that still serve an analytic purpose.[28] Following on from that observation, this book's argument has been that if scholars are interested in writing histories of the late nineteenth and early twentieth centuries that faithfully recapture the intentions and viewpoints of those who took part then we need to spend as much time examining the ideas that motivated them as we do the artefacts they left behind.

However, and connecting that emphasis on ideas with the points that have been made about the possible relationships between history and sociology, this book has also endeavoured to show how an intellectual history of science has great advantages when it comes to understanding how history continues to shape the ways we think and act in the present. As we have seen, Galton, Geddes, and Hobhouse had very different views of humans and their place in society and although Hobhouse might not be an explicit point of reference any more it is clear that his programme for British sociology provided a set of sensibilities for the discipline that subsequently developed. It is in this sense that sociologists, who view their history as a source of inspiration, have returned to Geddes in particular as they try to re-imagine their field in the early twenty-first century. There is no reason why that approach cannot be replicated elsewhere. History need not be about a past in which we simply preserve things that seem curious but ultimately unchallenging in light of our current experiences; it can be about a past that speaks directly to the present and is full of possibilities ready to be taken forward. It is in this sense that the history of British sociology told here is very much a history of futures past.

Notes

1 Introduction

1. Shils 1985: 166–7.
2. Huxley 1871: 152–3. Original emphasis.
3. Durkheim 1982: 162.
4. Mills 1959.
5. Durkheim 2005; Weber 1949.
6. Tönnies 2001.
7. Abbott 1999; Cristobal Young 2009. Demonstrating the extent to which they still command the attention of scholars in the early twenty-first century, a number of new studies of classical generation thinkers have been published in the past decade. For example, on Weber see Radkau 2009 and Ringer 2005, whilst on Durkheim see: Schmaus 2004 and Alexander and Smith 2005.
8. Weber 1958. On Weber's difficult relationship with German sociology, in particular the German Sociological Association, see: Proctor 1991: part 2, especially chapter 10.
9. For example: Giddens 1971. Indeed, interpreting the relationship between Marx and Weber is a crucial part of understanding how thinkers from different sides of the political spectrum, such as Talcott Parsons and Perry Anderson, who are discussed shortly, have given quite different readings of not just Weber but classical sociology as a whole.
10. For more on Parsons, see: Hamilton 1983; Holmwood 1996.
11. Parsons 1937. Parsons' translations of Weber were, however, conservative compared to those of Weber's European translators. As a consequence, the American reading of Weber, which presents him as providing an alternative to Marx's historical materialism, is frequently very different to European interpretations of Weber. For more on the mid-twentieth-century political context for these different readings see: Roy MacLeod 2006.
12. Martin Bulmer 1985: 15; Shils 1980: 182, n. 7; MacRae 1972: 47. Indeed, Tawney provided a foreword to Parsons' translation of Weber's *The Protestant Work Ethic and the Spirit of Capitalism*. See: Weber 1958. For more on the importance of Durkheim to British anthropology, see: Stocking 2001.
13. Parsons 1937: part 1. For a more recent take on this view of British sociological thought as a tradition beginning with Hobbes, see: Levine 1995: chapter 7.
14. Anderson 1968. As Lawrence Goldman has recently noted, it can be no coincidence that this article appeared in 1968 – a year when left-leaning British academics were caught up in a frenzy of excitement for the seemingly revolutionary potential of the intellectual culture of their European neighbours. Goldman 2007: 431–2. Indeed, following on from notes 9 and 11, it should be noted that Anderson's leftwing politics makes his reading of the classical canon far more radical than those offered by Parsons and other, mainly American, writers, including Edward Shils. For an excellent contextualization

of Anderson's essay in terms of British leftwing and Marxist politics of the 1960s, see Collini 2006: chapter 8. See also Roy MacLeod 2006.
15. Goldman 1983, 1987; Kent 1981; Abrams 1968.
16. For a reply to the normative claims of scholars such as Anderson and Parsons, see: Goldman 2007. See also: Roger Smith 1997: 535.
17. Abrams 1968; Hawthorn 1976: chapter 8. Furthermore, as Goldman has recently observed, one should not look down upon the close relationship between social scientists, social reformers, and government in nineteenth- and early twentieth-century Britain because it was actually the cause of much envy from Europe, where such integration was hoped for but not achieved. See: Goldman 2003, 2007: 434–5.
18. Collini 1978, 1979; den Otter 1996.
19. Lepenies 1988; Halsey 2004.
20. See for example: Abrams 1968; Collini 1979; den Otter 1996; Halsey 2004.
21. Barnes 1948a: 604.
22. Halliday 1968.
23. As the conclusion outlines in greater detail this historiographic stance owes much to the insights offered by Niall Ferguson (1998) and Gregory Radick, John Henry, Steve Fuller, Peter Bowler, and Steven French in their contributions to the recent focus section on counterfactual history in *Isis* (2008). For more on the problem of 'precursors' in the writing of history, see: Hodge 1990.
24. Shapin 2009, 2011.
25. Dear 2009.
26. Creath 2010; Fuller 2008, 2010; Erickson 2010a, 2010b; Fara 2010; Rouse 2010.
27. Secord 2004.
28. Golinski 1998: 9; Topham 2004: 432.

1   Political Economy, the BAAS, and Sociology

1. Ingram 1879: 642.
2. *The Times*, 23 August 1878, 10.
3. Abrams 1968: 177–95.
4. Koot 1987: 53–9.
5. Beatrice Webb 1926: 301–2, n.1.
6. Martin 1893: ii. 359. On Lowe's career see: Briggs 1972: 232–63; Bryce 1903; Martin 1893; Maloney 2005; Winter 1976.
7. Cairnes 1873: 232–3.
8. Bagehot 1876: 215.
9. This summary glosses over a whole range of complicated issues from the history of late eighteenth- and early nineteenth-century economic thought, including how much importance should be granted to Smith as *the* founder of political economy and whether classical political economists of the mid-nineteenth century actually continued a project outlined by Smith and then Ricardo. Regardless of these issues, the point remains that mid-nineteenth-century political economists saw themselves as working in the tradition of

Smith and Ricardo, even if Smith and Ricardo may not have approved of the way in which they had modified their ideas. For more on these issues see: Schumpeter 1976; Blaug 1958; Hutchison 1978; Schabas 2003, 2005; Backhouse 1985, 2002.

10. Backhouse 1985: 26; 2002: 136–7.
11. For useful summaries of classical theory see: Hutchison 1978: chapters 2 and 3; Schabas 2003.
12. Hobsbawn 1999: 211.
13. Collini 1991, especially chapter 1; Goldman 2002; Backhouse 2002: chapter 7; Stigler 1965. Though, as Margaret Schabas notes, if university chairs are to be taken as a measure of prestige, political economy was ahead of chemistry for most of the nineteenth century. Schabas 1997.
14. Political Economy Club 1882: 5. For more on the role of the Political Economy Club in British economics, see Henderson 1983; Tribe 2011.
15. Goldman 2002.
16. Robert Lowe, Speech to the House of Commons, 3 May 1865, 'Second Reading of the Borough Franchise Extension Bill', *Hansard Parliamentary Debates,* Third Series, Vol. 178, 21 March 1865– 8 May 1865, Col. 1426. Lowe's speeches on this topic were collected and printed with a preface in Lowe 1867.
17. Lowe, Speech to the House of Commons, 3 May 1865, Cols. 1431–2.
18. Lowe 1867: 15.
19. Lowe 1867: 10.
20. Lowe, Speech to the House of Commons, 3 May 1865, (op. cit. n. 16), Col. 1439.
21. Church 1975: 76–8; and Mitchell 1988: 524.
22. See: Inkster, Griffin, and Rowbottom 2000, for example.
23. On wages, see: Church 1975: 71–5. On poor relief figures see: Local Government Board 1901–2: 312; on assessing and interpreting those figures see: Aschrott 1888: 282, n.1, who claimed that the real numbers were treble the official five per cent estimate; Sidney and Beatrice Webb 1906–29: II.ii. 1041–3; and Best 1985: 153–68. See Perkins 2000 for a useful summary on these and related issues.
24. Foxwell 1887: 84.
25. Campbell 1877: 186.
26. Morrell and Thackray 1981: 291–6.
27. The source for historians' emphasis on Quetelet is Charles Babbage's accounts of events. See: Goldman 1983: 592–3.
28. Goldman 1983.
29. Morrell and Thackray 1981: 267.
30. On Whewell, Ricardian political economy and the BAAS, see: Goldman 1983; Henderson 1994; Maas 2005: 64–70; Porter 1994.
31. Richard Jones 1831: vii. The *Essay* on rent was intended to be the first of a four-part series. However, the other three parts (wages, profits, and taxation) were never published. For more on Jones and the content and context of historical arguments against Ricardian economics, see: Richard Jones 1831 and 1859; Koot 1987: 38–9; W. L. Miller 1971: 198–207; Backhouse 1985: 212–18.

32. On the foundation of the SSL see: Cullen 1975: 77–91; Goldman 1983; Hilts 1978; and Morrell and Thackray 1981: 291–6.
33. Whewell to Murchison, 25 September 1840, in Todhunter 1876: ii. 289. Also quoted in Howarth 1922: 87. On the Glasgow meeting see Morrell and Thackray 1981: 294–295; and Whewell's letter to Lord Northampton, 5 October 1840, in Todhunter 1876: ii. 292–4.
34. Quoted in Howarth 1922: 87, source not cited. A similar quote appears in: Whewell to Northampton, in Todhunter 1876: ii. 294.
35. Morrell and Thackray 1981: 296.
36. Henry Dunning Macleod 1872–5, 1874–5. See also Backhouse 1985: chapters 4 and 18.
37. Harrison 1865, 1870; Cairnes 1870a, 1870b. See also: Adelman 1971.
38. Harrison 1865: 356.
39. Sidgwick 1887: 4.
40. Bagehot 1876: 216.
41. Price 1879: 116.
42. Rogers 1888: vii.
43. Sidgwick 1887: 4–5; Mill 1869: 517.
44. Leslie 1876: 294.
45. For a range of assessments of the 'marginal revolution' see the special issue in the 1972 volume of the *History of Political Economy.* See in particular: Black 1972.
46. Jevons 1863a, 1863b. On marginalism before 1871 see, for example: de Marchi 1972. On sales figures: Black 2004: 103.
47. J. S. Mill, Speech to the House of Commons, 17 April 1866, 'Malt Duty Resolution', *Hansard Parliamentary Debates,* Third Series, Vol. 182, 12 March 1866–26 April 1866, Col. 1525.
48. Sidgwick 1887: 5, citing as evidence: Brassey 1873. See also Gareth Stedman Jones 1971.
49. Giffen 1879.
50. Mitchell 1988: 524.
51. Mitchell 1988: 257.
52. See: Gooday 2000.
53. Cairnes 1873: 241.
54. Jevons 1879: xv–xvi.
55. Jevons 1876: 618.
56. Political Economy Club of London 1876: 5.
57. Political Economy Club of London 1876: 21.
58. Jevons 1876: 619.
59. Jevons 1876: 620.
60. Council of the British Association for the Advancement of Science 1878: xlix. The official records of the BAAS contain very few conclusive traces of this committee. Francis Galton's participation in it is only confirmed by Farr's testimony in 'Proceedings of the Forty-Third Anniversary Meeting' 1877: 342.
61. Quoted in Farr's testimony in 'Proceedings of the Forty-Third Anniversary Meeting' 1877: 342–3. Original report not present in BAAS archives. The reference to a 'society [that] has been specially formed for the discussion of social and economical questions' was no doubt to the Social Science Association.

62. 'Council Meeting, Thursday 12th July 1877', Minute Book of the Council of the Statistical Society of London: 1873–89, Archives of the Royal Statistical Society, Royal Statistical Society, London, B2/3: 188.
63. 'Council Meeting, Thursday 15th November 1877', Minute Book of the Council of the Statistical Society of London: 1873–89, Archives of the Royal Statistical Society, B2/3: 195.
64. Council of the British Association for the Advancement of Science 1878: xlix.
65. Galton 1877b: 472.
66. Giffen and Chubb, quoted by Farr 1877: 473.
67. Farr 1877: 474.
68. Giffen and Chubb, quoted by Farr 1877: 475.
69. Galton 1877b: 471.
70. Galton 1877b: 471.
71. Farr 1877: 473. Original emphasis.
72. Giffen and Chubb, quoted by Farr 1877: 474.
73. Pearson 1914–24: ii. 348.
74. Giffen and Chubb, quoted by Farr 1877: 475.
75. Editorial, 'The Economic Section of the British Association', *Economist*, 8 September 1877, 1060–1; Anon. ('F.R.S'), 'The Proposed Discontinuance of Section F, Economic Science and Statistics, at the British Association', *Economist*, 20 October 1877, 1247–8, reprinted in 1877 in *Journal of the Statistical Society of London*, 40: 631–3.
76. 'Council Meeting, Thursday 11th July 1878', Minute Book of the Council of the Statistical Society of London: 1873–89, Archives of the Royal Statistical Society, B2/3: 222.
77. Council of the British Association for the Advancement of Science 1879: lvi.
78. Ingram 1879: 642.
79. Ingram 1879: 641.
80. Ingram 1879: 642.
81. Ingram 1879: 645.
82. Ingram 1879: 656.
83. Ingram 1879: 657.
84. Harrison 1865.
85. Minute Book of Section F, Papers of the British Association for the Advancement of Science, Bodleian Library, Oxford University, Dep. BAAS 334: 27.
86. Editorial, *The Times*, 23 August 1878, 10.
87. Lowe 1878.
88. Lowe 1878: 859.
89. Ingram 1879: 861.
90. Ingram 1879: 858.
91. For example, whilst Abrams includes a transcript of Ingram's address in his history of early British sociology, Backhouse provides a summary of Ingram's four-point critique of classical political economy in his history of modern economic analysis. Abrams 1968: 177–95; Backhouse 1985: 215.

2 Francis Galton and the Science of Eugenics

1. Galton 1877b: 471.
2. For an attempt to relate Galton to such trends see: Waller 2001.
3. Kevles 1985: chapter 1.
4. Donald A. Mackenzie 1981a; Cowan 1972a, 1977, 1985.
5. Pearson 1914–24; Forrest 1974; Michael Bulmer 2003; Gillham 2001.
6. Cowan 1977.
7. Galton 1855. Galton's turn to heredity is noted in his wife Louisa's diary, which is transcribed in Pearson 1914–24: ii. 70.
8. Galton 1909b: 287.
9. Galton 1909b: 288.
10. Galton 1909b: 289.
11. Galton 1865: 157.
12. Galton 1865: 165.
13. Galton 1865: 165.
14. Galton 1865: 165.
15. Donald A. Mackenzie 1981a: 52–6.
16. Galton 1909b: 288.
17. Galton 1869: 2. Diane B. Paul also notes that Galton's view that heredity underpinned social success and progress was not widely held at this time. See: Paul 2003: 229.
18. For more on how Victorian political economists grappled with questions about human behaviour that their science often raised, see Schabas 1997. For more on the history of 'economic man' and its problematic place in economic science see: Morgan 2006. For an extended analysis see: Davis 2003.
19. Galton 1869: 14.
20. Galton 1869: 14.
21. Indeed, Galton is seen by a number of scholars as being part of a larger group that helped initiate and shape a debate about human evolution and progress that Darwin later joined with his book of 1871, *The Descent of Man*. See: Paul 2003: 215–17; and Richards 1987: chapter 4.
22. On the historical background to and historiographical debate about this issue see: Gooday 2000.
23. Cowan 1977: 153–8. Cowan's brief but general argument is that Galton can be understood as being part of a trend of scepticism about the certainty of social progress amongst Victorian intellectuals, which has been noted by some historians of ideas. For example, see: Houghton 1957; Burn 1964.
24. Galton 1865: 166.
25. Galton 1873a: 116.
26. Galton 1873a: 120. An example of such forces were those encountered in the city, where Galton believed conditions were favourable to those in possession of less desirable qualities. This idea was alluded to by Galton in this article but explored more fully later in the year in: Galton 1873b.
27. Galton 1865: 157.
28. Galton 1869: 37.
29. Galton 1869: 10.
30. Galton 1909b: 304.

31. Quetelet's two main works of this period were *Sur l'homme et le développement de ses facultés, ou, Essai de physique sociale,* of 1835, and *Lettres sur les probabilities,* of 1845. Galton directed his readers to the 1849 English translation of the latter work: Galton 1869: 26. For the historical background and development of error theory, see: Hacking 1990; Donald A. MacKenzie 1981a; Stephen M. Stigler 1986; Gooday 2004: chapter 2. See also Victor Hilts' essay on the comparisons between Quetelet and Galton's work: Hilts 1973. We should also recall from chapter 1 that Quetelet's presence at the 1833 Cambridge meeting of the British Association for the Advancement of Science is frequently, though incorrectly, cited by many scholars as the reason for the formation of the Association's statistical section.

32. Galton 1869: 33.

33. Gökyigit 1994: 215–40

34. Pearson 1914–24: ii. 115.

35. Darwin 1868: ii. 357.

36. Darwin 1868: ii. chapter 27.

37. Darwin 1868: ii. 374.

38. For a long-term view of Darwin's ideas on this subject see: Hodge 1985.

39. Galton 1869: 373.

40. Galton 1869: 371–3.

41. Darwin 1868: ii. 374.

42. Emma Darwin to Henrietta Darwin, 19 March 1870, in Pearson 1914–24: ii. 158. See also Galton to Darwin, 12 May 1870, and Pearson's commentary in Pearson 1914–24: ii. 160.

43. Galton 1909b: 297.

44. Galton 1870–71: 404.

45. Darwin 1871: 502.

46. Galton 1871: 5. For more on Darwin's thought about inheritance see Hodge 1985. For more about the pangenesis dispute see: Gillham 2001: chapter 13; Pearson 1914–24: ii. chapter 10B; Cowan 1977: 164–79; Gayon 1998: 105–15; and Michael Bulmer 1999, though Bulmer's argument that Galton dropped his own theory of heredity because he realized that there were technical difficulties with it is refuted in the rest of this chapter.

47. See Galton and Darwin's correspondence, in Pearson 1914–24: ii. 166–77.

48. Galton 1875c: 81. As he often did throughout his career, Galton produced two different versions of this paper: one for a general audience and another, Galton 1876b, for a more specialist readership. See also: Galton 1871–2.

49. Indeed, whilst Galton mentioned pangenesis in *Natural Inheritance,* his landmark publication, he did not mention the stirp. The historiographic trend, begun by Pearson (1919–24: ii. 170–3), and most recently exemplified by Gillham (2001: 181–3), has been to present Galton's ideas in terms of their relationship to later orthodoxy. Galton's 'anticipation' of the continuity of the germ plasm, for example, is often justified by scholars through a letter of 1889 from August Weismann to Galton in which Weismann apologized for not having known earlier of Galton's work on the theory of heredity. Weismann to Galton, 23 February 1889, in Pearson 1914–24: iiia. 341.

50. Galton 1874b.

51. Galton 1875b: 35. The probable error, now known as the standard deviation, describes the spread of data.

52. Galton 1875b: 34.
53. Galton 1877a: 290. An excellent scene-setting description of occasion of this lecture is given in Gillham 2001: 199–205.
54. Galton 1877a: 290; 291. Galton's confidence on this point was motivated by the common belief that sweet peas are not subject to cross-fertilization, a notion that has subsequently been shown to be false.
55. Galton 1877a: 291.
56. Galton 1878.
57. Pearson 1914–24: ii. 283–4; Galton 1878: 97.
58. Indeed, although he never published on the subjects, Galton displayed an interest during the early and middle periods of his life in practices such as mesmerism and phrenology, which were then popular and are often portrayed as forerunners of modern psychology. See: Galton 1909b: 80; Gillham 2001: 215–16.
59. Galton 1879d: 141.
60. Galton 1879c. See also: Galton 1879b. See Galton 1909b: 259–65 for his reflections on the method, outcome and use of composite photography.
61. Galton 1879a: 425. See also: Galton 1879e.
62. Galton 1880. A revised version of the questionnaire appears in Galton 1883: Appendix E. For more on the questionnaires, including a transcript of the original see: Burbridge 1994. The presence of psychological studies in early volumes of *Mind*, which is now, of course, a philosophy journal, has been the subject of a number of recent studies. See: Green 2009; Staley 2009a and 2009b; Lanzoni 2009.
63. Galton 1880: 302.
64. For a historian of psychology's view of Galton, see: Fancher 1996: 216–45; as well as Sweeney 2001.
65. Galton 1883: 17, n. 1.
66. Galton 1883: 17, n. 1. Original emphasis.
67. Galton 1886a: 1207.
68. Galton 1874a, 1875a, 1876a.
69. Galton, 'Family Records', *The Times*, 9 January 1884, 10.
70. Galton 1884. The calculation of the worth of Galton's prizes in today's money comes from the formula and figures provided by O'Donoghue, Goulding, and Allen 2004: 41–3.
71. Galton 1882: 332.
72. Galton 1909b: 245.
73. As Raymond Fancher points out, a number of the seventeen measurements, including reaction times and head size, were rooted in Galton's belief that they were indicators of intellectual ability: Fancher 1996: 216–17. However, following the trend outlined at the end of the previous section, the anthropo-metric laboratory data did not enable Galton to substantiate such beliefs.
74. Galton, 'Prize Records of Family Faculties', *The Times*, 9 May 1884, 9; 'Prize Family Record', *The Times*, 27 June 1884, 12.
75. Galton 1885: 206.
76. Galton 1892b: 32.
77. Galton 1886a: 1206.
78. Galton 1886a: 1207.
79. Galton 1886a: 1210.

80. As was noted in the previous section, Galton had chosen sweet peas for his experiments in the mid-1870s because he believed that they were self-fertilizing – an assumption that has since been shown to be false.
81. Galton 1886a: 1208.
82. Galton 1886c: 254. Galton achieved this result by 'smoothing' his data – a common practice in which data is combined in ways that make it easier to understand and interpret. Gillham suggests that whilst Galton was on safe ground in this case he was subsequently guilty of over smoothing information to fit his expectations, specifically that about artistic faculty and tuberculosis in *Natural Inheritance*. See: Gillham 2001: 254; Galton 1889: chapters 9 and 10.
83. Galton 1886a: 1211.
84. Galton 1886a: 1211.
85. Galton 1886a: 1211. Dickson's mathematical solution was also published separately from Galton's work on heredity: Galton and Hamilton 1886.
86. Galton 1886a: 1211.
87. Galton 1886a: 1211.
88. Galton 1886b, 1886c, 1886d.
89. Patrick Geddes, 'Mr Francis Galton on Natural Inheritance', *Scottish Leader*, 14 March 1890, 2.
90. Galton 1883: 305.
91. Galton 1889: chapter 3.
92. Galton 1889: 28.
93. Galton 1889: 28. For further elaboration on and illustrations of this point see: Gayon 1998: 170–2; as well as Stephen Jay Gould's section on Galton's view of sports as the locus for evolutionary change in his mammoth final book: Gould 2002: 342–51.
94. Galton 1883: 334.
95. Bowler 1989: 64–73. See also: Cowan 1972b.
96. For background and further detail on the biometrician/Mendelian dispute see: Provine 1971: chapters 2 and 3; Gayon 1998: 197–319; Donald A. MacKenzie 1981b: 24–88; 1981a: chapters 5 and 6; Frogatt and Nevin 1971; Kevles 1981; Farrall 1975; Gillham 2001: chapters 19–21; de Marrais 1974.
97. Weldon 1894–95: 381.
98. See, for example, Pearson's critical appraisal of *Natural Inheritance* and specifically his analysis of Galton's use of regression: Pearson 1914–24: iiia. 57–79. On 'Galton's law of ancestral heredity' see Pearson 1897–98; Galton 1897. See also: Magnello 1998.
99. See: Donald A. MacKenzie 1981b on this point.
100. Bateson 1894.
101. Weldon 1893. Pearson published a series of papers in the *Proceedings of the Royal Society of London* between 1893 and 1904 detailing his statistical developments, which all appeared under the general title 'Mathematical Contributions to the Theory of Evolution'.
102. Galton 1894.
103. For more on this committee and its activities see: Frogatt and Nevin 1971: 10–12; Pearson 1914–24: iiia. 126–37.
104. Pearson 1914–24: iiia. 87–8.
105. Galton 1892a: 7.

3   Patrick Geddes' Biosocial Science of Civics

1. Patrick Geddes, 'Mr Francis Galton on Natural Inheritance', *Scottish Leader*, 14 March 1890, 2.
2. For example: Geddes 1904a, 1915.
3. Meller 1990; Welter 2002. The major biographical works on Geddes have been written by friends and former students of Geddes. Aside from Meller and Welter's work, most writing on Geddes has, as Alex Law (2005) has noted, tended to be of an unmistakably hagiographic nature. For example, see: Boardman 1944, 1978; Defries 1927; Kitchen 1975; Mairet 1957.
4. The exceptions to this rule are Renwick 2009; Radick and Gooday 2004.
5. Studholme 2007; Scott and Husbands 2007; Fuller 2007b; Studholme, Scott, and Husbands 2007. See also Studholme 2008; Fuller 2007a: 142–52.
6. See, for example, Stalley 1972. Geddes is frequently, though, incorrectly cited as the originator of the phrase 'think global, act local'. See, for example: Stephen 2004. Indeed, Geddes' leftwing stock rose during the mid to late twentieth century. The resurgence of interest in his work came primarily through his association with his greatest intellectual disciple – the American sociologist and architectural critic Lewis Mumford, whose writings, including *The Pentagon of Power* from his late 1960s Myth of the Machine series, were popular amongst American radicals during the Cold War. See: Renwick and Gunn 2008.
7. Robert M. Young 1970; Richards 1987. For a typical presentation of Spencer as an apologist for social Darwinism see: Gould 2000: 257–67.
8. Mayr 1982: 386.
9. Ruse 1996, 2004; Radick 2007: chapters 5 and 6. Indeed, in his final book, Gould backtracked somewhat on his earlier claims about Spencer when, in a discussion focused on an 1893 exchange between Spencer and August Weismann, Gould begrudgingly conceded that Spencer was someone who not only took part in serious biological debate but who also sometimes made points of scientific substance. See: Gould 2002: 197–208. On Spencer and scientific practice, see: Renwick 2009.
10. Dixon 2003, 2008; Beck 2003. See also Greta Jones 2003 for an analysis of how Spencer's rise to a position of immense popularity was linked to developments in publishing.
11. Spencer 1858. See also Spencer 1851. Spencer's early writings are most easily available in: Spencer 1868–74.
12. Spencer 1867: 285 (§ 97).
13. Spencer 1864–7: i. part V.
14. See in particular: Spencer 1864–7: i. 433–63 (§164–8).
15. Spencer 1887: 9.
16. Spencer 1864–7: i. 435 (§159).
17. Spencer 1864–7: ii. 508 (§ 377).
18. For an extensive discussion of this debate between Huxley and Spencer see: Elwick 2003. See also: Freeden 1978: 95.
19. Francis 2007: chapter 17.
20. Spencer 1870–2; Spencer 1873; Spencer 1879. For more on Spencer and altruism, see: Dixon 2003, 2008. For more on the military/industrial distinction,

see: Spencer 1896. On the end of the biological struggle for existence see, for example: Spencer 1864–7: ii. 507–8 (§ 377).

21. On Spencer and the political left, see: Beck 2003.
22. The exception is: Radick and Gooday 2004.
23. Huxley 1893–4a, 1893–4b. See also: Peel 1971: 151–3; Richards 1987: 313–19.
24. Quoted by Boardman 1978: 35. Source not cited. One can only speculate how it was that Geddes came to read Spencer at that exact moment in time but it seems more than a coincidence that it was whilst working with Michael Foster, who was a Spencer devotee. See: Kuklick 1998: 167; Gieson 1978: 353, n.
25. Geddes 1925: 742.
26. As Adrian Desmond has explained, despite Huxley's reputation as 'Darwin's Bulldog', he was a very different animal when teaching in his laboratory. Many of Darwin's ecological and population approaches lay outside the disciplinary norms of anatomy and thus they were not studied in Huxley's classroom. Indeed, Huxley frequently left his students to make up their own minds on most evolutionary questions, stressing only, of course, his objection to the links made by thinkers such as Spencer between natural and the social worlds. See: Desmond 1997: 70–1.
27. Boardman 1978: 35.
28. For Geddes' account of the importance to him of Huxley's early tutelage and patronage see: Geddes 1925.
29. Sapp 1994: 1–14.
30. Sapp 1994: 10. In this sense, the debate between Huxley and Spencer about the organization of individual organisms, which has been studied by Elwick, complements my examination here of 'symbiosis' between different organisms. Elwick 2003.
31. Geddes 1880–2b: 378. Geddes' time in Roscoff and, later Naples, was in keeping with the late nineteenth-century trend for biologists to attend sites such as marine stations for the purpose of developing their skills and furthering their research. See: Nyhart 1995.
32. Geddes 1878: 450. See also Geddes 1879.
33. Geddes 1878: 450.
34. Geddes 1878: 452.
35. Geddes 1878: 453.
36. Haeckel 1862: ii. For a recent study of the intellectual origins and inspiration for Haeckel's radiolarian research see: Richards 2004.
37. Geddes 1880–2b: 383.
38. Geddes 1880–2b: 387.
39. Geddes 1882: 303–5.
40. Moseley 1882: 338.
41. Brandt 1881.
42. Geddes 1880–2b: 391. Geddes' comments on Brandt's research are contained in a postscript Geddes added to his Ellis Physiology Prize winning paper in the *Proceedings of the Royal Society of Edinburgh*.
43. Geddes 1880–2b: 392.
44. Charles Darwin to Patrick Geddes, 27 March 1882, Geddes Papers, National Library of Scotland, MS 10522, f. 11.

45. August Weismann, 'Letter of Recommendation', in Geddes 1888a: 91.
46. Geddes 1880: 255. In fact, it was during this period of convalescence that Geddes invented what he called 'thinking machines' – a diagrammatic style of representing the dynamic relationship between different objects of study, including ideas and artefacts, which he did not think could be grasped through words alone. Geddes attached a great deal of importance to thinking machines throughout his career (see chapter 6) but few other people seem to have grasped exactly what they meant. For more on thinking machines, including numerous reproductions, see: Boardman 1978: 4–84. It should be noted, though, that Lewis Mumford described one of Geddes' larger and most elaborate of thinking machines as 'in effect, an effort to replace Herbert Spencer's Synthetic Philosophy... by a condensed version, a graph of thirty-six squares'. Mumford: 1982: 157.
47. Geddes 1881, 1880–2a. Whilst Geddes 1881 is an abstract of the paper that he delivered, the only difference between the two presentations appears to be his explicit efforts in his presentation to Section F to position his argument in terms of Ingram's earlier address.
48. Geddes 1881: 526.
49. Geddes 1881 and 1880–2a are essentially the same paper, though the address to the Royal Society of Edinburgh contains more detail of Geddes' specific proposals.
50. Geddes 1881: 524; Hooper 1881. Hooper, too, used Ingram's address as the starting point for his analysis.
51. Geddes 1880–2a: 302; Geddes 1881: 524.
52. Geddes 1880–2a: 204. Geddes 1881: 524.
53. Geddes 1881: 524. Categories A-D were included in Geddes 1880–2a: 304. Category E was added by Geddes in his BAAS paper.
54. Geddes also argued that the fourth of Ingram's reforms – 'that economic laws and the practical prescriptions founded on these should be conceived and expressed in less absolute form' – would also have been satisfied in his paper to Section F if 'the limits of [his paper] permitted reference to generalisation and practice'. Geddes 1881: 526.
55. Geddes 1882–4b.
56. Geddes 1882–4b: 964, n. Emphasis in original.
57. See, for example, Geddes 1882–4a, 1884–6, 1886a; 1886b; 1888c.
58. For more on Thomson, see: Bowler 2005.
59. Geddes 1888c; Geddes and Thomson 1889a. There is some confusion as to whether the *Chambers'* articles were co-authored with J. Arthur Thomson or not. Boardman (1978: 67–8) attributes them solely to Geddes; Meller (1990: 80) to Geddes and Thomson. As with *The Evolution of Sex*, which is discussed shortly, the continuity with the themes and ideas of Geddes' earlier work but greater clarity in the writing strongly suggests co-authorship but with Thomson as the junior partner. It should also be noted that readers of Geddes and Thomson's work often felt able to distinguish between the two men's contributions. Indeed, Lewis Mumford remarked of his experience of reading Geddes and Thomson's book *Evolution* (1911) that he 'identified without difficulty the writer whose voice called [him]'. Whereas Thomson 'wrote in a supple English style', Mumford observed that Geddes' prose was 'more crabbed and cryptic'. '[Geddes'] was the

audacity of an original mind', Mumford wrote, 'never content blindly to follow established conventions, still less fashions of the moment'. Mumford 1982: 145.

60. Geddes and Thomson 1889a: 483.
61. Geddes 1888c: 85.
62. Geddes and Thomson 1889a: 484.
63. Geddes 1888c: 85.
64. Geddes and Thomson 1889a: 479.
65. Geddes 1888b: 9.
66. Geddes 1888b: 21, 8.
67. Geddes 1886a; Geddes 1886b. See also: Geddes: 1884–6; Geddes and Thomson 1884–6. Geddes and Thomson's collaboration on *The Evolution of Sex*, as discussed with regards to their other work, was one where Thomson's role was to help summarize existing scientific knowledge, help improve the clarity of writing when expressing Geddes's theories, and, as many exchanges in the *Geddes Papers* at the National Library of Scotland reveal, restrain Geddes's expansive scientific vision. On Thomson's role during the writing of *The Evolution of Sex*, see also: Boardman 1944: 120–1; Kitchen 1975: 96. A further sign of Geddes's guiding hand in writing of *The Evolution of Sex* was the book's frequent use of diagrams, which were clearly inspired by what he called 'thinking machines' (see note 46). Indeed, Karl Pearson, who was soon to fall under the spell of Francis Galton's statistical methods, felt that the use of diagrams in *The Evolution of Sex* was 'exaggerated': Pearson, annotation dated 8 November 1889, in University College London's copy of: Geddes and Thomson 1889: 314. This copy of the book is held by UCL in their main library stack.
68. Geddes and Thomson 1889b: 26; Geddes 1886b: 724. Indeed, according to the *Oxford English Dictionary*, the terms 'anabolic' and 'katabolic' were first used in 1876 by the Spencerian Michael Foster, under whom Geddes had been studying when he first read Spencer's *The Principles of Biology*.
69. Geddes and Thomson 1889b: 27.
70. Geddes and Thomson 1889b: 97–134.
71. Geddes and Thomson 1889b: 235.
72. Geddes and Thomson 1889b: 270.
73. Arnstein 1965; McLaren 1978; Royle 1980. For more on the relationship between contraception, eugenics and the 'woman question', see: Richardson 2003.
74. 'P. C. M' 1889–90: 531–2.
75. Geddes and Thomson 1889b: 297.
76. Geddes and Thomson 1889b: 293–8.
77. Geddes and Thomson 1889b: 267.
78. Dixon 2008: chapter 7; Conway 1973; Yeo: 1996: 197; Caird 1892: 827.
79. Geddes and Thomson 1889b: 267.
80. White, whose wealth came from the estate of his father, a successful linen merchant, had stipulated that Geddes should be required to teach at Dundee during the summer term only. For more on White see: Branford 1928; Husbands 2005.
81. For more on the problems of Edinburgh's Old Town during the nineteenth century, see: P. J. Smith 1980.

82. Hill 1998. Indeed, the 1880s, the decade when Booth began his surveys of the London poor, were a period when direct intervention became fashionable amongst social reformers. Canon Samuel Barnett, for example, had established Toynbee Hall, a university settlement in London's east end, where reform-minded graduates of Oxford University were encouraged to live for a period after their studies had ended so their skills could be used for the benefit of the poverty-stricken surrounding area. Famous residents of Toynbee Hall include William Beveridge. See: Meacham 1987.
83. See: Edinburgh Social Union 1885.
84. Mavor 1923: i. 215.
85. Mavor 1923: i. 216.
86. For more on Le Play, see: Brooke 1970; Fletcher 1969; Fletcher 1971: i. appendix 3; Porter 2011; Silver 1982. See also: Demolins 1901–3, from whom Geddes learned of Le Play.
87. Brooke 1970: 105–6; Livingstone 1992: 272; Fletcher 1969: 56–7; Porter 2011.
88. Zueblin 1899. Geddes and Company published a range of Geddes' pamphlets and also handled the first print-run of *City Development* (1904) after it had been turned down by the Dunfermline Carnegie Trust (see chapter 5).
89. Geddes 1899: 945–6. Original emphasis.
90. For more on Branford see: Scott and Husbands 2007.
91. According to Helen Meller, the value of the property portfolio taken over by the Town and Gown Association was £52,555, which is approximately £4,400,000 in early twenty-first century money. See: Meller 1990: 78. For the details of the calculation, see: O'Donoghue, Goulding, and Allen 2004: 41–3.
92. Anna Geddes to Lady Welby, 9 October 1902, Welby Collection, 1970-010/006, folder 3.

4   L. T. Hobhouse's Evolutionary Philosophy of Reform

1. Hobhouse 1904: 109–10.
2. Hobhouse 1904: 111.
3. Hobhouse 1911, 1994.
4. Clarke 1978; Freeden 1978. For more on Hobson see: Hobson 1938, 1988; Freeden 1973.
5. Collini 1979; Freeden 1978. See also Barnes 1948b.
6. Boakes 1984; Radick 2007: chapter 6.
7. Fuller 2006: chapter 5.
8. Radick 2007: 211–14.
9. Hobhouse 1913: xv. Hobhouse repeatedly cited the end of his undergraduate career and the start of his Merton fellowship as the point where he formulated the research plan discussed in this chapter. See, for example: Hobhouse 1901: vi. See also: Collini 1979: 148–50. Radick, however, cites a letter of 1886 as evidence for Hobhouse having conceived of the project earlier than those conservative estimates. Radick 2007: 213.
10. Letter quoted by J. A. Hobson in: Hobson 1931: 28.

11. Hobhouse 1913: xv.
12. Cunningham 1996: introduction and chapter 1. For more on the history of philosophy during this period, in particular the emergence of analytic philosophy, which now dominates the field in Britain and America, see: Cunningham 1996; Metz 1938; Passmore 1966.
13. den Otter 1996: especially chapter 1.
14. See, for example: Ritchie 1901. For a general discussion of the relationship between idealist philosophy and evolutionary thought see: den Otter 1996: chapter 3.
15. Quoted by Graham Wallas, Review of J. A. Hobson and Morris Ginsberg, *L. T. Hobhouse: His Life and Work*, in *New Statesman and Nation*, 25 April 1931, 326.
16. Russell 1914: 26. For this reason, as Jonathan Hodge has pointed out to me, it should not be considered an historical accident that the name chosen by Russell and Moore for their new philosophical project was the oppositional term to Spencer's defining work, the 'Synthetic Philosophy'.
17. Indeed, the political philosophy of the British idealists continues to be interest to contemporary political philosophers. See: Vincent and Plant 1984; Nicholson 1990.
18. den Otter 1996: especially chapter 5; Freeden 1978: 94–116.
19. For more on Green's political philosophy and activities see: Green 1901, 1884. For more on Green's political writings and activities, see: Vincent and Plant 1984; Nicholson 1990; as well as Brink 2003.
20. Collingwood 1978: 17. It should be noted, though, that idealism also led to disputes in political philosophy. Most famously, Hobhouse and Bernard Bosanquet had a serious falling out in the early twentieth century over the relationship between the state, the individual, and society. See: Bosanquet 1899; Hobhouse 1918; Collini 1976; den Otter 1996.
21. For more on Hobhouse's involvement with trades unions and other social reform groups during this period see: Collini 1978: chapter 2. On Toynbee Hall: Meacham 1987.
22. Hobhouse 1893. The book also included a preface by R. B. Haldane – a Liberal M. P. whose younger brother had been Hobhouse's demonstrator in the Museum Laboratory at Oxford.
23. Hobhouse 1893: 90.
24. Hobhouse 1893: 53.
25. Hobhouse 1893: 54.
26. Hobhouse 1893: ix.
27. Hobhouse 1896: 3.
28. J. S. Mackenzie 1896: 396. The term 'epistemology' was coined by the Scottish philosopher James Frederick Ferrier (1808–64) in 1854 but it was not widely used until later in the nineteenth century. See: Dixon: 2004.
29. Hobhouse to Arthur Sidgwick, 18 November 1896, *Guardian* Archives, John Rylands University Library, University of Manchester, MS 132/1.
30. Barbara Hammond, quoted by Hobson 1931: 36.
31. Hobhouse to Sidgwick, 18 November 1896, *Guardian* Archives, MS 132/1.
32. Sidgwick to C. P. Scott, 20 November 1896, *Guardian* Archives, MS 132/2.
33. Ayerst 1971; Hammond 1934; *C. P. Scott, 1846–1932: The Making of the Manchester Guardian* 1946.

34. Hobhouse 1946: 84.
35. Hobson 1931: 40. See Scott's recollections of Hobhouse's productivity: Scott 1931: 7.
36. Hobhouse 1907b. On the various factions that sprang up during the Boer War, see: Davey 1978; Koss 1973; Matthew 1973; Semmel 1960.
37. Scott 1931: 7; Emily Hobhouse 1984; Fry 1929. Indeed, the Boer war was one of the major issues that drove a wedge between Hobhouse and the Fabian socialists, whose ideas Hobhouse had subtly endorsed in *The Labour Movement*. In addition to citing Beatrice Webb's *Cooperative Movement* as a source of inspiration, Hobhouse had advocated several Fabian policies in *The Labour Movement*, including the 'national minimum': a term that covered not just wages but a general across-the-board standard that nobody should be allowed to fall below. Through the radical contacts he had built at Oxford, Hobhouse had become friends with Beatrice and Sidney Webb, something that is evidenced by the acknowledgement to Hobhouse in the Webb's *Industrial Democracy* (1897). However, Hobhouse never joined the Fabian Party and, after the publication of *Fabianism and Empire* in 1900, he was never able to bring himself to bury his differences with the wing of the organization represented by George Bernard Shaw, who had suggested that the Boer War, empire, and collectivism were compatible. See: Hobhouse 1907b; and Shaw 1900. See also: McBriar 1962: chapters 4 and 5.
38. Hobhouse 1898: 156. See also: Hobhouse 1899.
39. Hobhouse 1901: vi.
40. Hobhouse 1901: v.
41. Radick 2007: chapter 6. See also: Boakes 1984: 179–84; Hearnshaw 1964: 101–4.
42. Hobhouse 1901: 152. On these experiments, see also: Hobhouse: 1902.
43. Hobhouse 1901: 152–3.
44. Hobhouse 1901: 155.
45. For more on Hobhouse's criticisms of Thorndike's methods, see: Hobhouse 1901: chapter 7. See also, Radick 2007: 211–12.
46. See: Hobhouse 1901: chapter 10.
47. Hobhouse 1901: chapter 17.
48. Hobhouse 1901: 5.
49. Hobhouse 1901: 388.
50. Hobhouse 1901: 388. See Hobhouse 1901: chapter 16.
51. Hobhouse 1901: 402–3.
52. Hobhouse to Scott, 4 February 1901, *Guardian* Archives, MS 132/103. Original emphasis.
53. Hobhouse did continue to contribute regularly to the newspaper after 1902, though he did so far less frequently than when he was a full staff member. Moreover, Hobhouse would formally reconnect with the *Guardian* in 1911 when he became a director of the newspaper.
54. 'The Sociological Society – Its Origins and Aims' 1904: 284.
55. For more on national efficiency in British politics, see: Searle 1990. For more on degeneration, see: Soloway 1990.
56. Hobhouse, Review of Lester Ward's *Pure Sociology*, *Manchester Guardian*, 28 April 1903, 4.

57. Hobhouse 1904: 83–4. This book was based on a series of articles that Hobhouse had originally published in the *Speaker* two years earlier. See: Hobhouse 1901–2a-h; Hobhouse 1902–3a-e.
58. Hobhouse 1904: 97–8.
59. Hobhouse 1904: 101.
60. For more on Pearson's ideas about social development, see: Porter 2004. On national efficiency see: Searle 1990.
61. Pearson 1901: 19. This passage was cited by Hobhouse. See: Hobhouse 1904: 114–15.
62. Pearson 1901: 19; Hobhouse 1904: 114. In the footnote on page 115, Hobhouse examines the difficulty Pearson's argument had under the weight of his recognition of this fact.
63. Hobhouse 1904: 115.
64. Hobhouse 1904: 115–16.
65. Hobhouse 1904: 108.
66. Hobhouse 1904: 118.
67. Seth 1908: 375.
68. Hobhouse to Lady Victoria Welby, 19 October 1904, Lady Victoria Welby Collection, Clara Thomas Special Collections and Archives, York University, Toronto, MS 1970-010/006.
69. Hobhouse 1906: i. v.
70. Hobhouse 1906: i. 13–14.
71. Hobhouse 1906: i. 19–20.
72. Hobhouse 1906: i. 24–5.
73. Hobhouse 1906: i. 40.
74. Hobhouse 1906: ii. 279.
75. Hobhouse 1906: ii. 280.
76. Russell 1907: 207. Original emphasis.
77. Russell 1907: 209. For a recent analysis of the fact/value distinction see: Glackin 2008.
78. Hobhouse 1907a: 323.
79. Hobhouse 1907a: 324.
80. Hobhouse 1907a: 325.
81. Indeed, as Stefan Collini has argued, Hobhouse was quite capable of formulating his ethical and political ideals in terms more palatable to analytic philosophers but chose not to. Collini 1979: 231. As letters exchanged between Hobhouse and Lady Victoria Welby in late 1906 make clear, Hobhouse did turn his attention to Russell's work for a brief period after finishing *Morals in Evolution*. However, Hobhouse had to 'confess', he wrote to Lady Welby, that he was 'baffled' by Russell's *Principles of Mathematics* – one of the foundation stones of analytic philosophy. 'A good deal of it I simply can't understand', Hobhouse explained, 'and probably I should have to spend some years in the higher mathematics before it would be intelligible. Other parts seem to be of questionable value, & sometimes of doubtful logic. Can you really *prove* that two & two make four by deducing it from his definition of number?' Hobhouse to Lady Welby, 18 November 1906, Welby Collection, 1970-010/006. Original emphasis.
82. For more on the dispute between Hobhouse and Thomasson, see: Collini 1979: 93–5; Hobson 1931: 44–6.

5   The Origins and Growth of the Sociological Society

1. Victor Branford to Patrick Geddes, 12 October 1904, Geddes Papers, National Library of Scotland, Edinburgh, MS 10556, f. 58. Original emphasis.
2. The two major studies of the Sociological Society are: Halliday 1968; Abrams 1968. For the most recent example of a study that quickly passes over the Sociological Society's early years in favour of the events that followed, see: Halsey 2004.
3. For example: Studholme 1997; Levitas 2010.
4. Halliday 1968; Abrams 1968.
5. On the links between the Sociological Society and the Eugenics Education Society, see: Mazumdar 1992: chapter 1.
6. Scott and Husbands 2007; Studholme 2007. See also: Studholme 2008; Fuller 2007b; Studholme, Scott and Husbands 2007. See also Fuller's reinterpretation of Hobhouse in: Fuller 2006: 59–61, which was also discussed in chapter 4.
7. For more on Lady Welby's life see Petrilli 2004; MacDonald 1912; Welby 1929; Welby 1931. For more on significs see Welby 1903 and 1911. For more on the wider influence of significs see: Pietarinen 2009.
8. Victor Branford to Patrick Geddes, 2 July 1902, Geddes Papers, MS 10556, f. 39.
9. J. G. Bartholomew to Victor Branford, 2 March 1903, Geddes Papers, MS 10592, f. 48.
10. Geddes to Lady Victoria Welby, 10 April 1903, Lady Victoria Welby Collection, Clara Thomas Special Collections of Archives, York University, Toronto, MS 1970-010/006, folder 3.
11. Geddes to Welby, 25 April 1903, Welby Collection, MS 1970-010/006, folder 3.
12. 'The Sociological Society – It Origins and Aims' 1904: 283. For more on the COS, see: Mowat 1961; Lewis 1995; Harris 1989.
13. There are no records of who was on this first Preliminary Committee. However, based on the papers of Benjamin Kidd, who had recently written the first article on sociology for the *Encyclopaedia Britannica* (Kidd 1902), it seems White, Haddon, and Loch were also involved. See the Sociological Society agendas and circular letters in: Kidd Papers, Cambridge University Library, Add. 8069, Box 3, S156–S181.
14. Circular Letter from the Provisional Committee of the Sociological Society, 25 June 1903, Kidd Papers, Add. 8069, Box 3, S156.
15. Of course, the question prompted by White's willingness to fund the Sociological Society almost single-handedly is why he was not prepared to underwrite Geddes' proposed 'Edinburgh School of Sociology' or 'Scottish Society of Sociology'. Whilst there is little in the way of evidence to solve this conundrum, it seems plausible that White's long-term experience of providing money for the Outlook Tower during the 1890s had left him wary of backing Geddes' ideas with hard cash. For more on White see: Branford 1928; Husbands 2005.
16. J. M. White to Sir Arthur Rücker, 29 June 1903, reprinted in 'Meeting of 22nd July 1903', *University of London, Minutes of Senate, 1902–1903*, University of

London Archives, Senate House Library, University of London, MS ST2/2/19, Minute 1680.

17. For a full list of members, see: 'Meeting of 22nd July 1903', *University of London, Minutes of Senate, 1902–1903*, University of London Archives, MS ST2/2/19, Minute 1681.

18. 'Report by the Committee Appointed to Consider Mr Martin White's Offer to Endow the Study of Sociology', reprinted in 'Meeting of 18th November 1903', *University of London, Minutes of Senate, 1903–1904*, University of London Archives, MS ST2/2/20, Minute 273. Unlike the Sociological Society, the School was never an attempt to establish sociology as a discipline and by 1911 it had been absorbed into the departments the Society had had helped establish at the London School of Economics.

19. On this issue, see also: Fincham 1975: 32–3.

20. White to Rücker, 5 July 1903, reprinted in 'Meeting of 22nd July 1903', *University of London, Minutes of Senate, 1902–1903*, University of London Archives, MS ST2/2/19, Minute 1680.

21. 'The Sociological Society – Its Origins and Aims' 1904: 283.

22. 'Sociology in England' 1904: 65.

23. For the agendas of these meetings, see: Kidd Papers, Add. 8069, Box 3, S153–75.

24. Branford to Kidd, 21 January 1904, Kidd Papers, Add. 8069, Box 3, S173.

25. For more on the concerns about the fitness of Boer War recruits and related concerns about degeneration, see: Soloway 1990: chapter 3; Paul 1998: chapter 1; Mazumdar 1992: chapter 1; Kevles 1985: chapter 5.

26. Branford to Welby, 18 November 1903, Welby Collection, MS 1970-010/002, folder 1. Galton's letter appears not to have survived. Though Branford told Welby that Galton had given £3 to the Society, the letter from Branford, acting as Sociological Society secretary, to Galton acknowledging the donation cites the amount as £5. See: Branford to Galton, 9 November 1903, Galton Papers, University College London, box 318.

27. Branford to Welby, 21 January 1904, Welby Collection, MS 1970-010/002, folder 1.

28. Branford to Galton, 9 November 1903, Galton Papers, box 318.

29. On the decline of the NAPSS, see: Goldman 2002.

30. Galton to Welby, 20 March 1904, Welby Collection, MS 1970-010/005, folder 6. See also: Galton to Welby, 8 April 1904, Welby Collection, MS 1970-010/005, folder 6.

31. Branford to Welby, 9 May 1904, Welby Collection, MS 1970-010/002, folder 2. Branford reported to Welby that this had been Pearson's reaction to the suggestion that eugenics be discussed at the Society without Galton.

32. Branford to Galton, 11 April 1904, Galton Papers, box 138/9A.

33. Galton to Welby, 11 April 1904, Welby Collection, MS 1970-010/005, folder 6. Original emphasis.

34. Welby to Galton, 12 April 1904, Welby Collection, MS 1970-010/005, folder 6.

35. Branford to Welby, 12 April 1904, Welby Collection, MS 1970-010/002, folder 2.

36. Galton to Welby, 12 April 1904, Welby Collection, MS 1970-010/005, folder 6.

37. This meeting also heard a paper from the Finnish anthropologist Edward Westermarck: Westermarck 1904.
38. Bryce 1904: xiv.
39. Galton 1904. This paper was subsequently published in *Nature* and included in Galton's compilation of essays: Galton 1909a: chapter 2.
40. Galton 1904: 45.
41. Galton 1904: 46.
42. Galton 1904: 50.
43. Galton 1904: 48.
44. Galton 1904: 49.
45. Pearson in 'Discussion – Eugenics: Its Definition, Scope and Aims' 1904: 52.
46. Indeed, so strong was this message that Branford felt compelled to write a lengthy response to Pearson, which was published in the first volume of *Sociological Papers*. See: Branford: 1904b.
47. Maudsley in 'Discussion – Eugenics: Its Definition, Scope and Aims' 1904: 54.
48. Weldon in 'Discussion – Eugenics: Its Definition, Scope and Aims' 1904: 56.
49. Wells in 'Discussion – Eugenics: Its Definition, Scope and Aims' 1904: 59.
50. Drysdale-Vickery in 'Discussion – Eugenics: Its Definition, Scope and Aims' 1904: 60. See also: Warner in 'Discussion – Eugenics: Its Definition, Scope and Aims' 1904: 56; Welby in 'Discussion – Eugenics: Its Definition, Scope and Aims' 1904: 76–8. For more on these kinds of arguments about contraception and female emancipation see: Richardson 2003; as well as the discussion of Geddes and J. Arthur Thomson's *The Evolution of Sex* in chapter 3.
51. Hobhouse in 'Discussion – Eugenics: Its Definition, Scope and Aims' 1904: 63.
52. Galton in 'Discussion – Eugenics: Its Definition, Scope and Aims' 1904: 78.
53. Branford to Galton, 12 May 1904, Galton Papers, box 318. Given the events to which it refers, the date on this letter is clearly incorrect. Indeed, Branford refers to the meeting of 'yesterday', which means that the letter was written on the 17 May.
54. Branford to Welby, 22 May 1904, Welby Collection, MS 1970-010/002, folder 2.
55. Branford to Welby, 22 May 1904, Welby Collection, MS 1970-010/002, folder 2.
56. Branford to Welby, 27 June 1904, Welby Collection, MS 1970-010/002, folder 2.
57. Branford to Welby, 24 May 1904, Welby Collection, MS 1970-010/002, folder 2; Branford 1904c; Durkheim 1904.
58. Indeed, contrary to received views of British sociology during this period as an enterprise that was uninformed about work being done overseas, Ferdinand Tönnies was a well-known and respected thinker in late nineteenth- and early twentieth-century Britain and was awarded the 1899 Welby Prize, which was endowed by Lady Welby and awarded by the journal *Mind*. See: Tönnies 1899. The Welby Prize was intended for essays that dealt with the issue of words and their meaning – Lady Welby's area of specialization in philosophy. See: Welby 1903; Welby 1911.

59. Geddes to Welby, 11 November 1903, Welby Papers, 1970-010/006, folder 3. Original emphasis.
60. Geddes to Welby, 4 January 1904, Welby Papers, 1970-010/006, folder 3.
61. J. H. Whitehouse to Geddes, 2 February 1904, Geddes Papers, MS10536, f. 8.
62. Geddes 1904a. The title page of *City Development* described Geddes as the 'President of the Edinburgh School of Sociology'.
63. Geddes 1904b: 105.
64. Geddes 1904b: 107; 108.
65. Geddes 1904b: 112.
66. Geddes 1904b: 113.
67. Geddes 1904b: 104. All papers at the Sociological Society were circulated in advance of the meetings. Geddes' digressions were included in the published version of his paper but printed in a different typeface to indicate the fact they were not part of the original manuscript. See chapter 3 for more on Geddes' reading of Galton.
68. Geddes 1904b: 111. J. M. Robertson in 'Discussion – Civics: As Applied Sociology' 1904: 122.
69. Geddes 1904b: 113.
70. Branford to Welby, 22 May 1904, Welby Collection, MS 1970-010/002, folder 2.
71. Geddes 1904b: 113.
72. Welby to Branford, 14 July 1904, Welby Collection, MS 1970-010/002, folder 2.
73. Amongst the terms Welby criticized were 'Culture Institute' and 'Culture Policy', neither of which she felt were necessary. Welby to Geddes, 2 August 1904, Welby Collection, MS 1970-101/006, folder 3. In addition to these terms, *City Development* also saw Geddes use the terms 'Paleotechnic' and 'Neotechnic' for the first time. Geddes use of these terms will be explored in chapter 6.
74. Welby to Geddes, 2 August 1904, Welby Collection, MS 1970-101/006, folder 3.
75. Haddon to Geddes, 21 August 1904, Geddes Papers, MS 10538, f. 88.

6   The End of Biological Sociology in Britain

1. Victor Branford to Lady Victoria Welby, 15 January 1907, Lady Victoria Welby Collection, Clara Thomas Special Collections and Archives, York University, Toronto, MS 1970-010/002, folder 8.
2. Scott and Husbands 2007; Studholme 2007; Fuller 2007b; Studholme, Scott and Husbands 2007.
3. Collini 1979: 209.
4. Branford 1929: 275; Boardman 1978: 231. See also: Mumford 1948: 681. Branford stated that Hobhouse's name 'was selected from four names submitted … for the permanent chair of sociology founded at the University of London'. As we will see, the four names of which Branford wrote were not all submitted for the Martin White chair but instead for posts within the 'Sociological Faculty'.

5. Arthur Geddes, interview with Patrick Boardman, cited in Boardman 1978: 231.
6. Of course, one may ask why there was an interview story at all. Whilst we may never get to the bottom of this question, it is plausible to suggest that the story was constructed, possibly by Geddes himself, from the events at the Sociological Society between 1903 and 1906. Perhaps Geddes considered the various presentations he gave during those years as an 'interview' of sorts? However, it is also possible that the story was a complete fabrication that was meant to explain why Geddes failed to capitalize on the advantages he had in the early twentieth-century British debate about sociology.
7. Branford to Lady Welby, 27 June 1904, Welby Collection, MS 1970-010/002, folder 2. Branford accepted the invitation from *Nature* himself. See: Branford 1904d. Branford later described the publication of this article as a 'signal advance for the [sociological] movement'. Branford to Welby, 14 September 1904, Welby Collection, MS 1970-010/002, folder 3.
8. Branford to Welby, 26 August 1904, Welby Collection, MS 1970-010/002, folder 3.
9. Galton to Branford, 13 October 1904, Geddes Papers, National Library of Scotland, Edinburgh, MS 10556, f. 60. Original emphasis. For records of Galton's offer to the University of London see: 'Meeting of Monday October 10th 1904', *Academic Council Minutes 1904–1905*, University of London Archives, Senate House Library, University of London, AC 1/1/5, Minutes 53–55; 'Meeting of Monday 17th October 1904', *Academic Council Minutes 1904–1905*, Minutes 74–6.
10. Branford to Geddes, 12 October 1904, Geddes Papers, MS 10556, f. 59.
11. Geddes to Branford, 15 October 1904, Geddes Papers, MS 10556, f. 62. Indeed, it should be noted that Branford, in a letter to Welby about the problems with the discussion after Galton's first presentation to the Sociological Society in May 1904, described Thomson as 'an ardent Galtonian', as compared with Geddes, who was, Branford wrote, 'a Galtonian', albeit a 'critical one'. Branford to Welby, 22 May 1904, Welby Collection, MS 1970-010/002, folder 2.
12. Geddes to Welby, 22 October 1904, Welby Collection, MS 1970-010/006, folder 3.
13. Branford to Geddes, 28 October 1904, Geddes Papers, MS 10556, f. 66.
14. Branford, 'Communication from the Sociological Society with reference to National Eugenics', Meeting of the Academic Council, Monday 14 and 28 November 1904, *Academic Council Minutes 1904–1905*, University of London Archives, AC 1/1/5, Minute 216.
15. Galton 1905a.
16. Galton 1905b.
17. It is worth noting that the German philosopher and sociologist Ferdinand Tönnies, author of *Germeinschaft und Gesellschaft* and a correspondent of Lady Victoria Welby, sent a written communication to the Galton session. Indeed, Tönnies wrote to Lady Welby in July 1904 to tell her of how he was 'following with keen attention the progress of the [Sociological Society], the foundation of which means an event to me. How much I am interested in Mr. Galton's theories and enquiries, you will recognise from the fact, that as early as 1882 I planned to take my domicile for a year or more on a small

island, in order to make a careful study of all the families living there with a view of discovering rules or caprices of heredity or caprices of heredity and more in particular of the effects of interbreeding (marriages of near kin) among mankind. I regret that I have omitted to execute this purpose, but I always lacked self-confidence, and also support and encouragement from others'. Tönnies to Welby, 12 July 1904, Welby Collection, MS 1970-010/018, folder 2.

18. Geddes to Welby, 22 October 1904, Welby Collection, MS 1970-010/006, folder 3.
19. Geddes 1905: 57.
20. Geddes 1905: 59.
21. Geddes 1905: 70.
22. Geddes 1905: 58.
23. Geddes 1905: 67–70.
24. Geddes 1905: 73.
25. Geddes 1905: 92.
26. See: Geddes 1904a: 174–5.
27. Geddes 1905: 107. Original emphasis.
28. Charles Booth in 'Discussion – Civics as Concrete and Applied Sociology, Part II' 1905: 112.
29. A. W. Stills in 'Discussion – Civics as Concrete and Applied Sociology, Part II' 1905: 115.
30. Rev. Dr. Aveling in 'Discussion – Civics as Concrete and Applied Sociology, Part II' 1905: 114.
31. E. S. Weymouth in 'Discussion – Civics as Concrete and Applied Sociology, Part II' 1905: 117.
32. Geddes to Welby, 23 August 1905, Welby Collection, MS 1970-010/006, folder 4.
33. Welby to Galton, 10 April 1905, Welby Collection, MS 1970-010/005, folder 7.
34. Sociological Society 1906: 13. For more on Slaughter, who would later serve as the first chairman of the organizing committee of the Eugenics Education Society, see: Mazumdar 1992: 276, n. 78; Soloway 1990: 56. For more on the relationship between the Eugenics Education Society and the Sociological Society, see the conclusion and: Mazumdar 1992: chapter 1.
35. Meeting of Monday June 26 1905, *Academic Council Minutes 1904–1905*, University of London Archives, MS AC 1/1/5, minutes 1033–1038.
36. Hobhouse in 'Discussion – Some Guiding Principles in the Philosophy of History by J. H. Bridges' 1905: 223.
37. See the Sociological Society newsletter of the 10 October 1905, in which the council 'beg[ged] to commend the object of the [Register of Able Families] to Members' and encouraged them to assist Galton and Schuster by filling out the questionnaires and documents that they had circulated: Galton Papers, University College London, Box 138/9A.
38. 'The Sociological Society – Its Origins and Aims' 1904: 283.
39. Reid 1906; McDougall 1906.
40. Thomson 1906.
41. Crawley 1906; Beatrice Webb 1906; Beveridge 1906. No further details are given in the *Sociological Papers* about the conference to which Beveridge presented his paper.

42. Geddes 1906.
43. Welby to Branford, 22 May 1906, Welby Collection, MS 1970-010/002, folder 8. Indeed, within only two months of his third presentation to the Sociological Society, Geddes' friends, including Branford, had started to discuss the possibility of setting up a 'Geddes Lectureship Fund' (see: Branford to Welby, 21 May 1906, Welby Collection, MS 1970-010/002, folder 8). Whilst the need for this fund was driven by Geddes' financial difficulties, the timing of the idea's appearance in correspondence, such as that between Branford and Welby, highlights that Geddes, or, at the very least, his friends, had lost all hope by mid-1906 of him securing a job in London.
44. Wells 1906: 364.
45. Wells 1906: 359.
46. Wells 1906: 365. For more on the general context of this notion of sociology as literature see the introduction; Lepenies 1988; Halsey 2004. Whilst, as Lepenies notes, figures such as Wells showed an interest in reconciling humanistic and scientific styles of social thought and, as Halsey has argued, that such concerns can be traced in the years after Hobhouse's appointment, Wells' presentation was the first occasion that a literary interpretation of sociology had been offered to the Sociological Society or entered its deliberations.
47. Meeting of Academic Council of the University of London, 25 June 1906, *Academic Council Minutes 1905–1906*, University of London Archives, AC 1/1/6, minutes 1073–1080. For more on the expansion of sociology in university teaching during this period, see: Fincham 1975.
48. Branford to Welby, 15 January 1907, Welby Collection, MS 1970-010/002, folder 8.
49. Hobhouse in 'Discussion – Eugenics: Its Definition, Scope and Aims'1904: 62–3. See also chapter 5.
50. Branford to Welby, 15 January 1907, Welby Collection, MS 1970-010/002, folder 8.
51. For more on Hartog, see: Hartog 1949.
52. Branford to Welby, 22 January 1907, Welby Collection, MS 1970-010/002, folder 8. Why Hartog would suggest a change of host institution is not clear. It is possible that, as a former colleague, he realized that Hobhouse was unlikely to react positively to the UCL idea. However, it is also possible that Hartog thought the University of London would be just as unlikely to approve of the plan. After all, when they had accepted White's offer to fund the teaching of sociology in 1903, the University of London had been very clear about the need for sociology to be differentiated from the courses that were already on offer to students (see chapter 5). In this sense, Hartog may have advised Branford that his plan was not sufficiently new to win the university's approval.
53. Branford to James Bryce, 6 February 1907, Welby Collection, MS 1970-010/002, folder 8.
54. Branford to Lady Welby, 8 February 1907, Welby Collection, MS 1970-010/002, folder 8.
55. Calculation made using formula in: O'Donoghue, Goulding, and Allen 2004: 41–3.
56. Branford's response to this need to exclude Slaughter was to attempt to find him a job elsewhere in London. See: Branford to Welby, 25 April 1907, Welby Collection, MS 1970-010/002, folder 8, in which Branford talks

about trying to secure for Slaughter the post of Lecturer in Experimental Psychology at UCL, which had become available after the resignation of William McDougall, who had given a paper entitled 'A Practicable Eugenics Suggestion' to the Sociological Society during the 1905–6 academic year. Although the job was 'a small affair', Branford wrote to Lady Welby, 'it might be just sufficient to keep Slaughter amongst us'.

57. James Martin White to Sir Arthur Rücker, 5 June 1907, reprinted in 'Meeting of 24th July 1907', *University of London, Minutes of Senate, 1906–1907*, University of London Archives, ST2/2/23, minute 2666. By 'Sociological Committee', White was referring to the Martin White Benefaction Committee.

58. 'Meeting of 24th July 1907', *University of London, Minutes of Senate, 1906–1907*, University of London Archives, ST2/2/23, minutes 2668 and 2670. This offer constituted a promotion for Westermarck from the lecturing position that White had funded at the University of London in 1906. After objections from the University of Helsingfors, where Westermarck held the chair in psychology, he had yet to take up the lecturing post.

59. 'Meeting of 24th July 1907', *University of London, Minutes of Senate, 1906–1907*, University of London Archives, ST2/2/23, minute 2669.

60. 'Meeting of 24th July 1907', *University of London, Minutes of Senate, 1906–1907*, University of London Archives, ST2/2/23, minute 2675.

Conclusion

1. Hobhouse 1908a: 7.
2. Hobhouse 1908a: 10.
3. Hobhouse 1908a: 18.
4. Hobhouse 1908a: 15.
5. Hobhouse 1908a: 15–16.
6. Hobhouse 1908a: 17–18.
7. Hobhouse 1908b: 9–10.
8. Famously, of course, Robert W. Fogel was awarded the Nobel Prize for economics in 1993, jointly with Douglass North, for a counterfactual investigation of the role of railroads in economic growth. For another important social science counterfactual study, see: Hawthorn 1991.
9. Ferguson 1998.
10. Radick 2008; Henry 2008; Bowler 2008; French 2008; Fuller 2008.
11. Mazumdar 1992; Soloway 1990.
12. For more on Le Play House see: Evans 1986.
13. Mumford 1940: 497; Geddes and Mumford 1995; Renwick and Gunn 2008; Gunn 2008.
14. Renwick and Gunn 2008. See also: Gunn 2008.
15. Fuller 2007b.
16. Geddes and Thomson 1931. For more on Geddes' and Thomson's popular science writings see: Bowler 2005.
17. Mumford 1955: 111.
18. According to Mumford's biographer D. L. Miller, who is amongst the few to have examined the manuscript in question, Mumford spent the final years of his life working on a project that was focused on the role of purpose in

evolution. Indeed, in a letter of December 1979, which is cited by Miller, Mumford wrote that 'if I were beginning life all over again I think I'd dedicate myself wholly to biology' and deal with the issues of chance, purpose, and causality better 'than anyone has yet done' – a comment that makes more sense when we recall that Mumford's first contact with Geddes' work was through his reading of Geddes and Thomson's 1911 book *Evolution*. See: D. L. Miller 1989: 558.

19. For obvious reasons, the claim that Geddes would have endorsed the Nazi party platform of the 1930s is a controversial and emotive one, especially as Fuller has done little to unpack the complex issues that his comments have raised. In one sense, it is important, as sociologists such as Zigmunt Bauman have argued, that we do not think that terms such as 'evil' explain the Nazi party and their policies, which were rooted in a range of scientific, technological, and philosophical developments that were utilized by other people and states, in particular the USA, during the period between World War I and World War II. Indeed, it is important to recognize, as Fuller has argued elsewhere, that the Holocaust was not where the Nazis began but where they ended and that, in this sense, we need to take seriously the idea that people who would not have approved of the Holocaust would certainly have approved of their policies of 1920s and 1930s. See: Fuller 2007a: 152–8; 2007b. See also: Fuller 2006: chapter 14. However, it does seem unlikely Geddes would have associated himself with the Nazi project. For one thing, questions have to be raised about whether Geddes' defiantly anti-statist political sensibilities would have permitted him to approve of the Nazi party. Secondly, we have to ask if his deeply moral outlook would have led him to endorse the Nazi agenda, even the relatively conservative one of the early 1930s. Indeed, Mumford, whose work throws much light on the Geddesian enterprise, wrote a book, entitled *Men Must Act*, in which he offered an impassioned attack on fascism in Europe, a defence of democracy, and called for US intervention rather than appeasement. See: Mumford 1939. See also: Mumford 1941.

20. Barnes 1948a: 603.

21. Wilson 1975. For more on the sociobiology debate, see: Segerstråle 2000; Alcock 2001; Kitcher 1985; Ruse 1979. For critical examinations of evolutionary psychology, see: Rose and Rose 2001; McKinnon 2006. See also: Dupré 2001. For more on the challenges for sociology posed by evolutionary biology see Jackson and Rees 2007.

22. For example, see: Runciman 1998. See also other works that explore the links between Darwinism and social theory, such as: Dickens 2000.

23. Runciman 2008.

24. For more on Ginsberg, see: Halsey 2004: 56–8. For more on the UNESCO statement and British science, see: Schaffer 2008: chapter 4.

25. For more on the history of sociology in Britain since Hobhouse's death in 1929, see: Martin Bulmer 1985; Halsey 2004; Halsey and Runciman 2005; Platt 2003. See also the articles on the history of twentieth-century British sociology in the recent issue of the *Sociological Review*, especially: Osborne, Rose, and Savage 2008a; 2008b.

26. Runciman 2005: 9–10. See also Halsey 2005: 16.

27. See: Fuller 2006: chapter 5.

28. Hamlin 2005.

Bibliography

Abbott, Andrew (1999). *Department and Discipline: Chicago Sociology at One Hundred.* Chicago: University of Chicago Press.

Abrams, Philip (1968). *The Origins of British Sociology: 1834–1914.* Chicago: University of Chicago Press.

Adelman, Paul (1971). 'Frederic Harrison and the Positivist Attack on Orthodox Political Economy', *History of Political Economy,* 3: 170–89.

Alcock, John (2001). *The Triumph of Sociobiology.* Oxford: Oxford University Press.

Alexander, Jeffrey C. and Phillip Smith (eds) (2005). *The Cambridge Companion to Darwin.* Cambridge: Cambridge University Press.

Anderson, Perry (1968). 'Components of the National Culture', *New Left Review,* 50: 3–57.

Arnstein, Walter L. (1965). *The Bradlaugh Case: A Study in Late Victorian Opinion and Politics.* Oxford: Clarendon Press.

Aschrott, Paul Felix (1888). *The English Poor Law System: Past and Present,* trans. H. Preston Thomas. London: Knight and Co.

Ayerst, David (1971). Guardian: *Biography of a Newspaper.* London: Collins.

Backhouse, Roger (1985). *A History of Modern Economic Analysis.* Oxford: Blackwell.

—— (2002). *The Penguin History of Economics.* London: Penguin.

Bagehot, Walter (1876). 'The Postulates of English Political Economy. No. '1', *Fortnightly Review,*19: 215–42.

Barnes, Harry Elmer (1948a). 'Introductory Note: English Sociology Since Spencer', in Harry Elmer Barnes (ed.), *An Introduction to the History of Sociology,* pp. 603-5. Chicago: University of Chicago Press.

—— (1948b). 'Leonard Trelawney Hobhouse: Evolutionary Philosophy in the Service of Democracy and Social Reform', in Harry Elmer Barnes (ed.), *An Introduction to the History of Sociology,* pp. 614–53. Chicago: University of Chicago Press.

Bateson, William (1894). *Materials for the Study of Variation Treated With Especial Regard to Discontinuity in the Origin of Species.* London: Macmillan & Co.

Beck, Naomi (2003). 'The Diffusion of Spencerianism and its Political Interpretations in France and Italy', in Greta Jones and Robert Peel (eds), *Herbert Spencer: The Intellectual Legacy,* pp. 37–60. London: Galton Institute.

Best, Geoffrey (1985). *Mid-Victorian Britain, 1851–75.* London: Fontana.

Beveridge, W. H. (1906). 'The Problem of the Unemployed', *Sociological Papers,* 3: 323–31.

Black, R.D. Collison (1972). 'W.S. Jevons and the Foundation of Modern Economics', *History of Political Economy,* 4: 364-78.

—— (2004). 'Jevons, William Stanley (1835–1882)', in *Oxford Dictionary of National Biography,* online ed., edited by Lawrence Goldman. Oxford: Oxford University Press, http://www.oxforddnb.com/view/article/14809 (accessed 20 June 2011).

Blaug, Mark (1958). *Ricardian Economics: A Historical Study.* New Haven: Yale Press.

Boakes, Robert A. (1984). *From Darwin to Behaviourism: Psychology and the Minds of Animals.* Cambridge: Cambridge University Press.

Boardman, Philip (1944). *Patrick Geddes: Maker of the Future.* Chapel Hill: University of North Carolina Press.

—— (1978). *The Worlds of Patrick Geddes: Biologist, Town Planner, Re-Educator, Peace-Warrior.* London: Routledge and Keegan-Paul.

Bosanquet, Bernard (1899). *The Philosophical Theory of the State.* London: Macmillan &Co.

Bowler, Peter J. (1989). *The Mendelian Revolution: The Emergence of Hereditarian Concepts in Modern Science and Society.* London: Athlone.

—— (2005). 'From Science to the Popularisation of Science: The Career of J. Arthur Thomson', in David Knight and Matthew D. Eddy (eds), *Science and Beliefs: From Natural Philosophy to Natural Science, 1700-1900*, pp. 231–48. Aldershot: Ashgate.

—— (2008). 'What Darwin Disturbed: The Biology That Might Have Been', *Isis* 99: 560–7.

Brandt, Karl (1881). 'Über das Zusammenleben von Algen und Tieren', *Biologisches Centralblatt*, 1: 524–7.

Branford, Victor (1904a). 'On the Origin and Use of the Word Sociology', *Sociological Papers*, 1: 1–24.

—— (1904b). 'Note on the History of Sociology in Reply to Professor Karl Pearson', *Sociological Papers*, 1: 25–42.

—— (1904c). 'On the Relation of Sociology to the Social Sciences and to Philosophy (Abstract of Paper)', *Sociological Papers*, 1: 200–3.

—— (1904d). 'Social Types and Social Evolution. Review of *Aspects of Social Evolution. First Series. Temperaments* by J. Lionel Tayler', *Nature*, 70: 449–50.

—— (1928). 'James Martin White', *Sociological Review*, 20: 340–1.

—— (1929). 'The Sociological Work of Leonard Hobhouse', *Sociological Review*, 21: 273–80.

Brassey, Thomas (1873). 'Work and Wages', *Edinburgh Review*, 138: 334–66.

Briggs, Asa (1972). *Victorian People: A Reassessment of Persons and Themes, 1851–67.* Chicago: University of Chicago Press.

Brink, David O. (2003). 'Editor's Introduction', in T.H. Green, *Prolegomena to Ethics*, new edn, David O. Brink (ed.), pp. xiii–cx. Oxford: Clarendon.

Brooke, Michael Z. (1970). *Le Play: Engineer and Social Scientist. The Life and Work of Frédéric Le Play.* London: Longman.

Bryce, James (1903). *Studies in Contemporary Biography.* London: Macmillan & Co.

—— (1904). 'Introductory Address', *Sociological Papers*, 1: xiii–xviii.

Bulmer, Martin (1985). 'The Development of Sociology and of Empirical Social Research in Britain', in Martin Bulmer (ed.), *Essays on the History of British Sociological Research*, pp. 3–36. Cambridge: Cambridge University Press.

Bulmer, Michael (1999). 'The Development of Francis Galton's Ideas on the Mechanism of Heredity', *Journal of the History of Biology*, 32: 263–92.

—— (2003). *Francis Galton: Pioneer of Heredity and Biometry.* London: Johns Hopkins University Press.

Burbridge, David (1994). 'Galton's 100: An Exploration of Francis Galton's Imagery Studies', *British Journal for the History of the Science*, 27: 443–63.

Burn, W. L. (1964). *The Age of Equipoise: A Study of the Mid-Victorian Generation.* London: George Allen & Unwin.

Caird, Mona (1892). 'A Defence of the So-Called "Wild Women"', *Nineteenth Century*, 31: 811–29.

Cairnes, J. E. (1870a). 'M. Comte and Political Economy', *Fortnightly Review*, 7: 579–602

—— (1870b). 'A Note', *Fortnightly Review*, 8: 246–8.

—— (1873). *Essays in Political Economy: Theoretical and Applied.* London: Macmillan & Co.

Campbell, George (1877). 'Presidential Address', *Report of the Forty-Sixth Meeting of the British Association for the Advancement of Science; Held at Glasgow in September 1876*, Section F—Economic Science and Statistics: 186–94.

Church, R. A. (1975). *The Great Victorian Boom, 1850-1873.* London: Macmillan Press.

Clarke, Peter (1978). *Liberals and Social Democrats.* Cambridge: Cambridge University Press.

Collingwood, R.G. (1978). *An Autobiography.* Oxford: Clarendon Press. (1st edn published 1939).

Collini, Stefan (1976). 'Hobhouse, Boasanquet and the State', *Past and Present*, 72: 86–111.

—— (1978). 'Sociology and Idealism in Britain 1880-1920', *Archives Européennes de Sociologie*, 19: 3–50.

—— (1979). *Liberalism and Sociology: L. T. Hobhouse and Political Argument in England, 1880-1914.* Cambridge: Cambridge University Press.

—— (1991). *Public Moralists: Political Thought and Intellectual Life in Britain, 1850–1930.* Oxford: Clarendon Press.

—— (2006). *Absent Minds: Intellectuals in Britain.* Oxford: Oxford University Press.

Conway, Jill (1973). 'Stereotypes of Femininity in a Theory of Sexual Evolution', in Martha Vicinus (ed.), *Suffer and Be Still: Women in the Victorian Age*, pp. 140–54. Bloomington: Indiana University Press.

Council of the British Association for the Advancement of Science (1878). 'Report of the Council for the Year 1876-77, presented to the General Committee at Plymouth on Wednesday, August 15th, 1877', *Report of the Forty-Seventh Meeting of the British Association for the Advancement of Science; Held at Plymouth in August 1877*: xlix-l.

—— (1879). 'Report of the Council for the Year 1877-78, presented to the General Committee at Dublin on Wednesday, August 14, 1878', *Report of the Forty-Eighth Meeting of the British Association for the Advancement of Science; Held at Dublin in August 1878*: lvi-lvii.

Cowan, Ruth Schwartz (1972a). 'Francis Galton's Statistical Ideas: The Influence of Eugenics', *Isis*, 63: 509–28.

—— (1972b). 'Francis Galton's Contribution to Genetics', *Journal of the History of Biology*, 5: 389–412.

—— (1977). 'Nature and Nurture: The Interplay of Biology and Politics in the Work of Francis Galton', *Studies in History of Biology*, 1: 133–208.

—— (1985). *Sir Francis Galton and the Study of Heredity in the Nineteenth Century.* New York: Garland Publishing.

C. P. Scott, 1846-1932: The Making of the Manchester Guardian (London: Muller, 1946).

Crawley, A. E. (1906). 'The Origin and Function of Religion', *Sociological Papers*, 3: 243–9.

Creath, Richard (2010). 'The Role of History in Science', *Journal of the History of Biology*, 43: 207–14.

Cullen, M.J. (1975). *The Statistical Movement in Early Victorian Britain: The Foundations of Empirical Research*. New York: Harvester Press.

Cunningham, Suzanne (1996). *Philosophy and the Darwinian Legacy*. Rochester: University of Rochester Press.

Darwin, Charles (1868). *The Variation of Plants and Animals Under Domestication*, 2 Vols. London: John Murray.

—— (1871). 'Pangenesis', *Nature*, 3: 502–3.

Davey, Arthur (1978). *The British Pro-Boers, 1877-1902*. Cape Town: Tafelberg.

Davis, John B. (2003). *The Theory of the Individual in Economics: Identity and Value*. London: Routledge.

Dear, Peter (2009). 'The History of Science and the History of the Sciences: George Sarton, *Isis*, and the Two Cultures', *Isis*, 100: 89–93.

Defries, Amelia (1927). *The Interpreter: Geddes. The Man and His Gospel*. London: Routledge.

De Marchi, N. B. (1972). 'Mill and Cairnes and the Emergence of Marginalism in England', *History of Political Economy*, 4: 344–63.

De Marrais, Robert (1974). 'The Double-Edged Effect of Sir Francis Galton: A Search for the Motives in the Biometrician-Mendelian Debate', *Journal of the History of Biology*, 7: 141–74.

Demolins, Edmond (1901-3). *Comment La Route Crée Le Type Sociale*, 2 Vols. Paris: Firmin-Didot

Den Otter, Sandra M. (1996). *British Idealism and Social Explanation: A Study in Late Victorian Thought*. Oxford: Clarendon Press.

Desmond, Adrian (1997). *Huxley: Evolution's High Priest*. London: Michael Joseph.

Dickens, Peter (2000). *Social Darwinism: Linking Evolutionary Thought to Social Theory*. Buckingham: Open University Press.

'Discussion—Civics: As Applied Sociology' (1904). *Sociological Papers*, 1: 119–29.

'Discussion—Civics as Concrete and Applied Sociology, Part II' (1905). *Sociological Papers*, 2: 112–19.

'Discussion—Eugenics: Its Definition, Scope and Aims' (1904). *Sociological Papers*, 1: 53–63.

'Discussion—Some Guiding Principles in the Philosophy of History by J.H. Bridges' (1905). *Sociological Papers*, 2: 221–32.

Dixon, Thomas (2003). 'Herbert Spencer and Altruism: the Sternness and Kindness of a Victorian Moralist', in Greta Jones and Robert Peel (eds), *Herbert Spencer: The Intellectual Legacy*, pp. 85–124. London: Galton Institute.

—— (2004). 'James Frederick Ferrier (1808-64)', in Bernard Lightman (ed.), *The Dictionary of Nineteenth-Century British Scientists*, 4 Vols., ii. 681–2. Bristol: Thoemmes.

—— (2008) *The Invention of Altruism: Making Moral Meanings in Victorian Britain*. Oxford: Oxford University Press for the British Academy.

Dupré, John (2001). *Human Nature and the Limits of Science*. Oxford: Clarendon.

Durkheim, Emile (1904). 'On the Relation of Sociology to the Social Sciences and to Philosophy (Abstract of Paper)', *Sociological Papers*, 1: 197–200.

—— (1982). *The Rules of Sociological Method*, (ed.) Steven Lukes, trans. W.D. Halls. New York. Free Press. (1st published 1895).

—— (2005). *Suicide: A Study in Sociology*, trans. John A. Spaulding and George Simpson, (ed.) and intro. George Simpson. London and New York: Routledge. (1st published 1897).

Edinburgh Social Union (1885). *Description of the Aims of the Society*. Edinburgh: Edinburgh Social Union.

Elwick, James (2003). 'Herbert Spencer and the Disunity of the Social Organism', *History of Science*, 41: 35–72.

Erickson, Mark (2010a). 'Why Should I Read Histories of Science?', *History of the Human Sciences*, 23: 68–91.

—— (2010b). 'Why Should I Read Histories of Science? A Response to Patricia Fara, Steve Fuller, and Joseph Rouse', *History of the Human Sciences*, 23: 105–8.

Evans, D. F. T. (1986). 'Le Play House and the Regional Survey Movement in British Sociology, 1920-55', Unpublished PhD Thesis, City of Birmingham Polytechnic.

Fancher, Raymond E. (1996). *Pioneers of Psychology*, 3rd edn. London: Norton. (1st edn published 1979).

Fara, Patricia (2010). 'Why Mark Erickson Should Read Different Histories of Science', *History of the Human Sciences*, 23: 92–4.

Farr, William (1877). 'Considerations, in the form of a Draft Report, submitted to Committee, favourable to the maintenance of Section F', in 'Economic Science and the British Association', *Journal of the Statistical Society of London*, 40: 473–6.

Farrall, Lyndsay Andrew (1975). 'Conflict and Controversy in Science: A Case Study—The English Biometric School and Mendel's Laws', *Social Studies of Science*, 5: 269–301.

Ferguson, Niall (1998). 'Introduction Virtual History: Towards a "Chaotic" Theory of the Past', in Niall Ferguson (ed.), *Virtual History: Alternatives and Counterfactuals*, pp. 1–90. New York: Basic.

Fincham, Jill (1975). 'The Development of Sociology First Degree Courses at English Universities, 1907-72', Unpublished PhD Thesis, City University, London.

Fletcher, Ronald (1969). 'Frédéric Le Play', in Timothy Raison (ed.), *The Founding Fathers of Social Science*, pp. 51–8. Harmondsworth: Penguin.

—— (1971). *The Making of Sociology: A Study of Sociological Theory*, 2 Vols. London: Michael Joseph.

Forrest, D.W. (1974). *Francis Galton: The Life and Work of a Victorian Genius*. London: Paul Elek.

Foxwell, H.S. (1887). 'Economic Movement in England', *Quarterly Journal of Economics*, 2: 84–103.

Francis, Mark (2007). *Herbert Spencer and the Invention of Modern Life*. Stocksfield: Acumen.

Freeden, Michael (1973). 'J.A. Hobson as a New Liberal Theorist: Some Aspects of his Social Thought Until 1914', *Journal of the History of Ideas*, 34: 421-43.

—— (1978). *The New Liberalism: An Ideology of Social Reform*. Oxford: Clarendon Press.

French, Steven (2008). 'Genuine Possibilities in the Scientific Past and How to Spot Them', *Isis*, 99: 568–75.

Frogatt, P., and N. C. Nevin (1971). 'Galton's "Law of Ancestral Heredity": Its Influence on the Early Development of Human Genetics', *History of Science*, 10: 1–27.

Fry, A. R. (1929). *Emily Hobhouse: A Memoir*. London: Jonathan Cape.

Fuller, Steve (2006). *The New Sociological Imagination*. London: Sage.

—— (2007a). *Science Vs Religion? Intelligent Design and the Problem of Evolution*. Cambridge: Polity.

—— (2007b). 'A Path Better Not to Have Been Taken', *Sociological Review*, 55: 807–15.

—— (2008). 'The Normative Turn: Counterfactuals and a Philosophical Historiography', *Isis*, 99: 576–84.

—— (2010). 'History of Science for its Own Sake?', *History of the Human Sciences*, 23: 95–9.

Galton, Francis (1855). *The Art of Travel: Or, Shifts and Contrivances Available in Wild Countries*. London: John Murray.

—— (1865). 'Hereditary Talent and Character', *MacMillan's Magazine*, 12: 157–66, 318–27.

—— (1869). *Hereditary Genius*. London: Macmillan & Co.

—— (1870–1). 'Experiments in Pangenesis, by Breeding from Rabbits of a Pure Variety, into Whose Circulation Blood Taken from other Varieties Had Previously Been Largely Transfused', *Proceedings of the Royal Society of London*, 19: 393–410.

—— (1871). 'Pangenesis', *Nature*, 4: 5.

—— (1871–2). 'On Blood-Relationship', *Proceedings of the Royal Society of London*, 20: 394–402.

—— (1873a). 'Hereditary Improvement', *Fraser's Magazine*, 7: 116–30.

—— (1873b). 'The Relative Supplies from Town and Country Families, to the Population of Future Generations', *Journal of the Statistical Society of London*, 36: 19–26.

—— (1874a). 'Proposal to Apply for Anthropological Statistics from Schools', *Journal of the Anthropological Institute of Great Britain and Ireland*, 3: 308–11.

—— (1874b). 'On a Proposed Statistical Scale', *Nature*, 9: 342–3.

—— (1875a). 'Notes on the Malborough School Statistics', *Journal of the Anthropological Institute of Great Britain and Ireland*, 4: 130–5.

—— (1875b). 'Statistics by Intercomparison, with Remarks on the Law of Frequency of Error', *The London, Edinburgh, and Dublin Philosophical Magazine and Journal of Science*, series 4, 49: 33–46.

—— (1875c). 'A Theory of Heredity', *Contemporary Review*, 27: 80–95.

—— (1876a). 'On the Height and Weight of Boys Aged 14, in Town and Country Public Schools', *Journal of the Anthropological Institute of Great Britain and Ireland*, 5: 174–81.

—— (1876b). 'A Theory of Heredity', *Journal of the Anthropological Institute of Great Britain and Ireland*, 5: 329–48.

—— (1877a). 'Typical Laws of Heredity', *Proceedings of the Royal Institution*, 8: 282–301.

—— (1877b). 'Considerations Adverse to the Maintenance of Section F (Economic Science and Statistics)', in 'Economic Science and the British Association', *Journal of the Statistical Society of London*, 40: 468–73.

—— (1878). 'Address to the Department of Anthropology', *Report of the Forty-Seventh Meeting of the British Association for the Advancement of Science; Held at Plymouth in August 1877*, Section H—Anthropology: 94–100.

—— (1879a). 'Psychometric Facts', *Nineteenth Century*, 5: 425–33.

—— (1879b). 'Generic Images', *Nineteenth Century*, 6: 157–69.

—— (1879c). 'Generic Images', *Proceedings of the Royal Institution*, 9: 161–70.

—— (1879d). 'Composite Portraits, Made by Combining Those of Many Different Persons Into a Single Resultant Figure', *Journal of the Anthropological Institute of Great Britain and Ireland*, 8: 132–44.

—— (1879e). 'Psychometric Experiments', *Brain*, 2: 149–62.

—— (1880). 'Statistics of Mental Imagery', *Mind*, 5: 301–18.

—— (1882). 'The Anthropometric Laboratory', *Fortnightly Review*, 31: 332–8.

—— (1883). *Inquiries into Human Faculty and Its Development*. London: Macmillan & Co.

—— (1884). *Record of Family Faculties*. London: Macmillan & Co.

—— (1885). 'On the Anthropometric Laboratory at the Late International Health Exhibition', *Journal of the Anthropological Institute of Great Britain and Ireland*, 14: 205–21.

—— (1886a). 'Presidential Address', *Report of the Fifty-Fifth Meeting of the British Association for the Advancement of Science; Held at Aberdeen in September 1885*, Section H—Anthropology: 1206–14.

—— (1886b). 'Family Likeness in Eye-Colour', *Proceedings of the Royal Society of London*, 40: 402–16.

—— (1886c). 'Regression Towards Mediocrity in Heredity Stature', *Journal of the Anthropological Institute of Great Britain and Ireland*, 15: 246–63.

—— (1886d). 'President's Address', *Journal of the Anthropological Institute of Great Britain and Ireland*, 15: 488–500.

—— (1889). *Natural Inheritance*. London: Macmillan & Co.

—— (1892a). 'Presidential Address', *Transactions of the Seventh International Congress of Hygiene and Demography*, 10: 7–12.

—— (1892b). 'Retrospect of Work Done at my Anthropometric Laboratory at South Kensington', *Journal of the Anthropological Institute of Great Britain and Ireland*, 21: 32–5.

—— (1894). 'Discontinuity in Evolution', *Mind*, 3: 362–72.

—— (1897). 'The Average Contribution of Each Several Ancestor to the Total Heritage of the Offspring', *Proceedings of the Royal Society of London*, 61: 401–13.

—— (1904). 'Eugenics: Its Definition, Scope and Aims', *Sociological Papers*, 1: 43–50.

—— (1905a). 'Restrictions in Marriage', *Sociological Papers*, 2: 3–13.

—— (1905b). 'Studies in National Eugenics', *Sociological Papers*, 2: 14–17.

—— (1909a). *Essays in Eugenics*. London: Eugenics Education Society.

—— (1909b) *Memories of My Life*, 3rd edn. London: Methuen.

Galton, Francis and J. D. Hamilton (1886). 'Family Likeness in Stature', *Proceedings of the Royal Society of London*, 40: 42–73.

Gayon, Jean (1998). *Darwinism's Struggle for Survival: Heredity and the Hypothesis of Natural Selection*. Cambridge: Cambridge University Press.

Geddes, Patrick (1878). 'Observations on the Physiology and Histology of Convoluta Schultzii', *Proceedings of the Royal Society of London*, 28: 449–57.

—— (1879). 'Chlorophylle animale et la physiologie des planaires vertes', *Archives de Zoologie Expérimentale et Générale*, 8: 51–8.

—— (1880). 'Report of the Committee, consisting of Dr. Gamgee, Professor Schafer, Professor Allman, and Mr Geddes, for conducting Palaeontological and Zoological Researches in Mexico', *Report of the 50th Meeting of the British Association for the Advancement of Science; Held at Swansea in August and September 1880*, Reports on the State of Science: 254–7.

—— (1880–2a). 'On the Classification of Statistics and its Results', *Proceedings of the Royal Society of Edinburgh*, 11: 295–322.

—— (1880–2b). 'On the Nature and Function of the 'Yellow Cells' of Radiolarians and Coelenterates', *Proceedings of the Royal Society of Edinburgh*, 11: 377–96.

—— (1881). 'Economics and Statistics, Viewed from the Standpoint of the Preliminary Sciences', *Nature*, 24: 523–6.

—— (1882). 'Further Researches on Animals Containing Chlorophyll', *Nature*, 25: 303–5.

—— (1882–4a). 'A Restatement of the Cell Theory, with Applications to the Morphology, Classification, and Physiology of Protists, Plants and Animals. Together with an Hypothesis of Cell Structure, and an Hypothesis of Contractility', *Proceedings of the Royal Society of Edinburgh*, 12: 266–92.

—— (1882–4b). 'An Analysis of the Principles of Economics', *Proceedings of the Royal Society of Edinburgh*, 12: 943–80.

—— (1884–6). 'Theory of Growth, Reproduction, Sex, and Heredity', *Proceedings of the Royal Society of Edinburgh*, 13: 911–31.

—— (1886a). 'Reproduction', in T.S. Baynes and William Robertson Smith (eds), *Encyclopaedia Britannica*, 24 Vols, 9th edn, xx. 407–22. Edinburgh: A. & C. Black.

—— (1886b). 'Sex', in T.S. Baynes and William Robertson Smith (eds), *Encyclopaedia Britannica*, 24 Vols, 9th edn, xx. 720–4. Edinburgh: A. & C. Black.

—— (1888a). 'University of Edinburgh, Chair of Botany: Letter of Application, etc.', Privately Printed, National Library of Scotland.

—— (1888b). *Co-Operation Versus Socialism*. Manchester: Co-Operative Printing Society LTD.

—— (1888c). 'Variation and Selection', in T.S. Baynes and William Robertson Smith (eds), *Encyclopaedia Britannica*, 24 Vols, 9th edn, xxiv. 76–85. Edinburgh: A. & C. Black.

—— (1899). 'The Edinburgh Outlook Tower', *Report of the Sixty-Eighth Meeting of the British Association for the Advancement of Science Held at Bristol in September 1898*, Section E—Geography: 945–7.

—— (1904a). *City Development: A Study of Parks, Gardens, and Culture-Institutes. A Report to the Carnegie Dunfermline Trust*. Edinburgh and Birmingham: Geddes and Company and the Saint George Press.

—— (1904b). 'Civics as Applied Sociology', *Sociological Papers*, 1: 101–18.

—— (1905). 'Civics as Applied Sociology. Part II', *Sociological Papers*, 2: 55–111.

—— (1906). 'A Suggested Plan for a Civic Museum (or Civic Exhibition) and its Associated Studies', *Sociological Papers*, 3: 197–230.

—— (1915). *Cities in Evolution: An Introduction to the Town Planning Movement and to the Study of Civics*. London: Williams and Norgate.

—— (1925). 'Huxley as Teacher', *Nature*, 115: 740–743.

Geddes, Patrick and Lewis Mumford (1995). *Lewis Mumford and Patrick Geddes: The Correspondence*, (ed.) Frank G. Novack Jr. London: Routledge.

Geddes, Patrick and J. Arthur Thomson (1884–6). 'History and Theory of Spermatogenesis', *Proceedings of the Royal Society of Edinburgh*, 13: 803–23.

—— (1889a). 'Evolution', in *Chambers' Encyclopaedia*, 10 Vols., iv. 477–84. London and Edinburgh: W. & R. Chambers.

—— (1889b). *The Evolution of Sex*. London: Walter Scott.

—— (1911). *Evolution*. London: Williams & Norgate.

—— (1914). *Sex*. London: Williams & Norgate.

—— (1925). *Biology*. London: Williams & Norgate.

—— (1931). *Life: Outlines of General Biology*, 2 Vols. London: Williams & Norgate.

Giddens, Anthony (1971). *Capitalism and Modern Social Theory: An Analysis of the Writings of Marx, Durkheim and Max Weber*. Cambridge: Cambridge University Press.

Giffen, Robert (1879). 'On the Fall of Prices of Commodities in Recent Years', *Journal of the Statistical Society of London*, 42: 36–78.

Gieson, Gerald (1978). *Michael Foster and the Cambridge School of Physiology*. Princeton: Princeton University Press.

Gillham, Nicholas Wright (2001). *A Life of Sir Francis Galton: From African Exploration to the Birth of Eugenics*. New York: Oxford University Press.

Glackin, Shane Nicholas (2008). 'The Role of the Fact/Value Distinction in Modern Moral Life', Unpublished PhD Thesis, University of Leeds.

Gökyigit, Emel Aileen (1994). 'The Reception of Francis Galton's *Hereditary Genius* in the Victorian Periodical Press', *Journal of the History of Biology*, 27: 215–40.

Goldman, Lawrence (1983). 'The Origins of British "Social Science": Political Economy, Natural Science and Statistics, 1830–1835', *Historical Journal*, 26: 587–616.

—— (1987). 'A Peculiarity of the English? The Social Science Association and the Absence of Sociology in Nineteenth-Century Britain', *Past and Present*, 114: 133–71.

—— (2002). *Science, Reform, and Politics in Victorian Britain: The Social Science Association, 1857-1886*. Cambridge: Cambridge University Press.

—— (2003). 'Civil Society in Nineteenth-Century Britain and Germany: J.M. Ludlow, Lujo Brentano, and the Labour Question', in Jose Harris (ed.), *Civil Society in British History. Ideas, Identities, Institutions*, pp. 97–113. Oxford: Oxford University Press.

—— (2007). 'Foundations of British Sociology 1880-1930: Contexts and Biographies', *Sociological Review*, 55: 431–40.

Golinski, Jan (1998). *Making Natural Knowledge: Constructivism and the History of Science*. Cambridge: Cambridge University Press.

Gooday, Graeme (2000). 'Lies, Damned Lies and Declinism: Lyon Playfair, the Paris 1867 Exhibition and the Contested Rhetorics of Scientific Education and Industrial Performance', in Ian Inkster, Colin Griffin, and Judith Rowbottom (eds), *The Golden Age: Essays in British Social and Economic History, 1850-1870*, pp. 105–20. Aldershot: Ashgate.

—— (2004). *The Morals of Measurement: Accuracy, Irony, and Trust in Late Victorian Electrical Practice*. Cambridge: Cambrdige University Press.

Gould, Stephen Jay (2000). *The Lying Stones of Marrakech: Penultimate Reflections in Natural History*. London: Jonathan Cape.

—— (2002). *The Structure of Evolutionary Theory*. Cambridge, MA: Belknap Press of Harvard University Press.

Green, Christopher D. (2009). 'The Curious Rise and Fall of Experimental Psychology in *Mind*', *History of the Human Sciences*, 22: 37–57.

Green, T. H. (1884). *Prolegomena to Ethics*, 2nd edn, (ed.) A. C. Bradley. Oxford: Clarendon Press.

—— (1901). *Lectures on the Principles of Political Obligation*. London: Longmans, Green and Co.

Gunn, Richard C. (2008). 'A Critical Examination of Lewis Mumford's Account of Technics', Unpublished PhD Thesis, University of Leeds.

Hacking, Ian (1990). *The Taming of Chance*. Cambridge: Cambridge University Press.

Haeckel, Ernst (1862). *Die Radiolarien (Rhizopoda Radiaria)*, 2 vols. Berlin: Reimer.

Halliday, R. J. (1968). 'The Sociological Movement, the Sociological Society and the Genesis of Academic Sociology in Britain', *Sociological Review*, 16: 377–98.

Halsey, A. H. (2004). *A History of Sociology in Britain: Science, Literature and Society*. Oxford: Oxford University Press.

—— (2005). 'The History of Sociology in Britain', in A.H. Halsey and W.G. Runciman (eds), *British Sociology Seen from Without and Within*, pp. 13–22. Oxford: Oxford University Press for the British Academy.

Halsey, A. H. and W. G. Runciman (eds) (2005). *British Sociology Seen from Without and Within*. Oxford: Oxford University Press for the British Academy.

Hamilton, Peter (1983). *Talcott Parsons*. London: Tavistock.

Hamlin, Christopher (2005). 'Games Editors Played or Knowledge Readers Made?', *Isis*, 96: 633–42.

Hammond, J. L. (1934). *C. P. Scott of the* Manchester Guardian. London: G. Bell.

Harris, Jose (1989). 'The Webbs, the Charity Organisation Society and the Ratan Tata Foundation: Social Policy from the Perspective of 1912', in Martin Bulmer, Jane Lewis, and David Piachaud (eds), *The Goals of Social Policy*, pp. 27–63. London: Unwin Hyman.

Harrison, Frederic (1865). 'The Limits of Political Economy', *Fortnightly Review*, 1: 356–76.

—— (1870). 'Professor Cairnes on M. Comte and Political Economy', *Fortnightly Review*, 8: 39–58.

Hartog, Mabel (1949). *P. J. Hartog: A Memoir*. London: Constable.

Hawthorn, Geoffrey (1976). *Enlightenment and Despair: A History of Sociology*. Cambridge: Cambridge University Press.

—— (1991). *Plausible Worlds: Possibility and Understanding in History and the Social Sciences*. Cambridge: Cambridge University Press.

Hearnshaw, Leslie Stephen (1964). *A Short History of British Psychology, 1840-1940*. London: Methuen.

Henderson, James P. (1983). 'The Oral Tradition in British Economics: Influential Economists in the Political Economy Club of London', *History of Political Economy*, 15: 149–79.

—— (1994). 'The Place of Economics in the Hierarchy of the Sciences: Section F from Whewell to Edgeworth', in Philip Mirowski (ed.), *Natural Images in Economic Thought: 'Markets Read in Tooth and Claw'*, pp. 484–535. New York: Cambridge University Press.

Henry, John (2008). 'Ideology, Inevitability, and the Scientific Revolution', *Isis*, 99: 552–9.

Hill, Octavia (1998). *Octavia Hill and the Social Housing Debate: Essays and Letters by Octavia Hill*, (ed.) Robert Whelan. London: IEA Health and Welfare Unit.

Hilts, Victor L. (1973). 'Statistics and Social Science', in Ronald N. Giere and Richard S. Westfall (eds), *Foundations of Scientific Method: The Nineteenth Century*, pp. 206–33. Bloomington: Indiana University Press.

—— (1978). '*Alias exterendum*, or, the Origins of the Statistical Society of London', *Isis*, 69: 21–43.

Hobhouse, Emily (1984). *Boer War Letters*, Rykie Van Reenen (ed.). Cape Town: Human and Rousseau.

Hobhouse, L. T. (1893). *The Labour Movement*. London: T. Fisher Unwin.

—— (1896). *The Theory of Knowledge: A Contribution to Some Problems of Logic and Metaphysics*. London: Methuen and Co.

—— (1898). 'The Ethical Basis of Collectivism', *International Journal of Ethics*, 8: 137–56.

—— (1899). 'The Foreign Policy of Collectivism', *Economic Review*, 9: 197–220.

—— (1901). *Mind in Evolution*. London: Macmillan & Co.

—— (1901–2a). 'Some Shattered Illusions', *Speaker*, 5: 300.

—— (1901–2b). 'The Limitations of Democracy', *Speaker*, 5: 359–60.

—— (1901–2c). 'Democracy and Liberty', *Speaker*, 5: 388–9.

—— (1901–2d). 'Democracy and Nationality', *Speaker*, 5: 415–16.

—— (1901–2e). 'Democracy and Imperialism, *Speaker*, 5: 443–4.

—— (1901–2f). 'Democracy and Imperialism', *Speaker*, 5: 474–75.

—— (1901–2g). 'The Intellectual Reaction', *Speaker*, 5: 501–2.

—— (1901–2h). 'The Intellectual Reaction', *Speaker*, 5: 526–7.

—— (1902). 'The Diversions of a Psychologist', *Pilot*, 4 January 1902, 12–13; 11 January 1902, 36–7; 1 February 1902, 126–7; 1 March 1902, 232–3; 29 March 1902, 344–5; 26 April 1902, 449–51.

—— (1902–3a). 'Democracy and Empire', *Speaker*, 7: 75–6.

—— (1902–3b). 'Philosophy in England', *Speaker*, 7: 282–3.

—— (1902–3c). 'Sociology in America', *Speaker*, 7: 344–5.

—— (1902–3d). 'Henry Sidgwick on his Contemporaries', *Speaker*, 7: 421–2.

—— (1902–3e). 'The Laws of Hammurabi', *Speaker*, 7: 551–2.

—— (1904). *Democracy and Reaction*. London: T. Fisher Unwin.

—— (1906). *Morals in Evolution: A Study in Comparative Ethics*, 2 vols. London: Chapman and Hall.

—— (1907a). 'Sociology and Ethics', *Independent Review*, 12: 322–31.

—— (1907b). 'The Career of Fabianism', *Nation*, 1: 182–3.

—— (1908a). 'The Roots of Modern Sociology', in *Inauguration of the Martin White Professorships of Sociology, December 17th 1907*, pp. 7–23. London: John Murray.

—— (1908b). 'Editorial', *Sociological Review*, 1: 1–11.

—— (1911). *Liberalism*. London: Williams and Norgate.

—— (1913). *Development and Purpose: An Essay Towards A Philosophy of Evolution*. London: Macmillan & Co.

—— (1918). *The Metaphysical Theory of the State: A Criticism*. London: G. Allen and Unwin.

—— (1946). 'Liberal and Humanist', in *C. P. Scott, 1846-1932: The Making of the Manchester Guardian*, pp. 84–90. London: Muller.

—— (1994). *Liberalism and Other Writings*, James Meadowcroft (ed.). Cambridge: Cambridge University Press.

Hobsbawn, Eric (1999). *Industry and Empire: From 1750 to the Present Day*, new edn. London: Penguin.

Hobson, J. A. (1931). 'L.T. Hobhouse: A Memoir', in J.A. Hobson and Morris Ginsberg, *L. T. Hobhouse: His Life and Work*, pp. 15–95. London: George Allen and Unwin.

—— (1938). *Confessions of an Economic Heretic*. London: George Allen and Unwin.

—— (1988). *J. A. Hobson: A Reader*, Michael Freeden (ed.). London: Unwin Hyman.

Hodge, M. J. S. (1985). 'Darwin as a Lifelong Generation Theorist', in David Kohn (ed.), *The Darwinian Heritage*, pp. 207–43. Princeton: Princeton University Press.

—— (1990). 'Origins and Species: Before and After Darwin', in Robert Olby, G.N. Cantor, J.R.R. Christie, and M.J.S. Hodge (eds), *Companion to the History of Modern Science*, pp. 374–95. London: Routledge.

Holmwood, John (1996). *Founding Sociology? Talcott Parsons and the Idea of General Theory*. New York: Longman.

Hooper, Wynnard (1881). 'The Method of Statistical Analysis', *Journal of the Statistical Society of London*, 44: 31–49.

Houghton, Walter E. (1957). *The Victorian Frame of Mind, 1830-1870*. New Haven: Yale University Press.

Howarth, O. J. R. (1922). *The British Association for the Advancement of Science: A Retrospect, 1831-1921*. London: BAAS.

Husbands, Christopher T. (2005). 'James Martin White (1857–1928) as the Godfather of British Sociology?', *Sociology Research News: Newsletter of the London School of Economics and Political Science Sociology Department*, 3: 1–2.

Hutchison, T. W. (1978). *On Revolutions and Progress in Economic Knowledge*. Cambridge: Cambridge University Press.

Huxley, T. H. (1871). 'The Scientific Aspects of Positivism', in *Lay Sermons, Addresses and Reviews*, 3rd edn, pp. 147–73. London: Macmillan & Co. (Article originally published 1869).

—— (1893–4a). 'Administrative Nihilism', T.H. Huxley, *Collected Essays*, 9 Vols, i. pp. 251–89. London: Macmillan & Co. (Article originally published in 1871).

—— (1893–4b). 'The Struggle for Existence in Human Society', in T.H. Huxley, *Collected Essays*, 9 Vols, ix. pp. 195–236. London: Macmillan & Co. (Article originally published in 1887).

Ingram, J. K. (1879). 'Presidential Address', *Report of the Forty-Eighth Meeting of the British Association for the Advancement of Science; Held at Dublin in August 1878*, Section F—Economic Science and Statistics: 641–58.

Inkster, Ian, Colin Griffin, and Judith Rowbottom (eds) (2000), *The Golden Age: Essays in British Social and Economic History, 1850-1870*. Aldershot: Ashgate.

Jackson, Stevi and Amanda Rees (2007). 'The Appalling Appeal of Nature: The Popular Appeal of Evolutionary Psychology as a Problem for Sociology', *Sociology*, 41: 917–930.

Jevons, W. S. (1863a). 'The Study of Periodic Commercial Fluctuations', *Report of the Thirty-Second Meeting of the British Association of the Advancement of Science; Held at Cambridge in October 1862*, Section F—Economic Science and Statistics: 157–8.

—— (1863b). 'Notice of a General Mathematical Theory of Political Economy', *Report of the Thirty-Second Meeting of the British Association of the Advancement of Science; Held at Cambridge in October 1862*, Section F—Economic Science and Statistics: 158–9.

—— (1876). 'The Future of Political Economy', *Fortnightly Review*, 20: 617–31.

—— (1879). *The Theory of Political Economy*, 2nd edn. London: Macmillan & Co.

Jones, Gareth Stedman (1971). *Outcast London: A Study in the Relationship Between Classes in Victorian Society*. Oxford: Clarendon Press.

Jones, Greta (2003). 'Spencer and his Circle', in Greta Jones and Robert Peel (eds), *Herbert Spencer: The Intellectual Legacy*, pp. 1–16. London: Galton Institute.

Jones, Richard (1831). *An Essay on the Distribution of Wealth, and on the Sources of Taxation. Part One—Rent.* London: John Murray.

—— (1859). *Literary Remains, Consisting of Lectures and Tracts on Political Economy, of the Late Rev. Richard Jones*, William Whewell (ed.). London: John Murray.

Kent, Raymond A. (1981). *A History of British Empirical Sociology*. Aldershot: Gower.

Kevles, Daniel J. (1981). 'Genetics in the United States and Great Britain 1890–1930: A Review with Speculations', in Charles Webster (ed.), *Biology, Medicine and Society, 1840-1940*, pp. 193–216. Cambridge: Cambridge University Press.

—— (1985). *In the Name of Eugenics: Genetics and the Uses of Human Heredity*. New York: Alfred A. Knopf.

Kidd, Benjamin (1902). 'Sociology', in Donald Mackenzie Wallace, A.T. Hadley, and H. Chisholm (eds), *Encyclopaedia Britannica*, 35 vols., 10th edn, xxxii. 692–8. Edinburgh: A. & C. Black.

Kitchen, Paddy (1975). *A Most Unsettling Person: An Introduction to the Ideas and Life of Patrick Geddes*. London: Victor Gollancz.

Kitcher, Philip (1985). *Vaulting Ambition: Sociobiology and the Quest for Human Nature*. Cambridge, MA: The MIT Press.

Koot, Gerard M. (1987). *English Historical Economics, 1870-1926. The Rise of Economic History and Neomercantalism*. New York: Cambridge University Press.

Koss, Stephen (ed.) (1973). *The Pro-Boers: The Anatomy of an Anti-War Movement*. Chicago: University of Chicago Press.

Kuklick, Henrika (1998). 'Fieldworkers and Physiologists', in Anita Herle and Sandra Rouse (eds), *Cambridge and the Torres Strait: Centenary Essays on the 1898 Anthropological Expedition*, pp. 158–80. Cambridge: Cambridge University Press.

Lanzoni, Susan, (2009). 'Sympathy in *Mind* (1876-1900)', *Journal of the History of Ideas*, 70: 265–87.

Law, Alex (2005). 'The Ghost of Patrick Geddes: Civics as Applied Sociology', *Sociological Research Online*, 10. <http://www.socresonline.org.uk/10/02/law.html>

Lepenies, Wolf (1988). *Between Literature and Science: The Rise of Sociology*, trans. R. J. Hollingdale. Cambridge: Cambridge University Press.

Leslie, Thomas Edward Cliffe (1876). 'On the Philosophical Method of Political Economy', *Hermathena*, 4: 265–96.

Levine, Donald N. (1995). *Visions of the Sociological Tradition*. Chicago: University of Chicago Press.

Levitas, Ruth (2010). 'Back to the Future: Wells, Sociology, Utopia and Method', *The Sociological Review*, 58: 530–47.

Lewis, Jane E. (1995). *The Voluntary Sector, the State, and Social Work in Britain: The Charity Organisation Society/Family Welfare Association Since 1869*. Aldershot: Elgar.

Livingstone, David N. (1992). *The Geographical Tradition: Episodes in the History of a Contested Enterprise*. Oxford: Blackwell.

Local Government Board (1901–2). *Thirty-First Annual Report*. London: HMSO.

Lowe, Robert (1867). *Speeches and Letters on Reform*, 2nd edn. London: Robert John Bush.

—— (1878). 'Recent Attacks on Political Economy', *Nineteenth Century*, 4: 858–68.

Maas, Harro (2005). *William Stanley Jevons and the Making of Modern Economics*. New York: Cambridge University Press.

McBriar, A. M. (1962). *Fabian Socialism and English Politics, 1884-1918*. Cambridge: Cambridge University Press.

MacDonald, W. (1912). 'Lady Welby', *Sociological Review*, 5: 152–6.

McDougall, W. (1906). 'A Practicable Eugenic Suggestion', *Sociological Papers*, 3: 53–80.

MacKenzie, Donald A. (1981a). *Statistics in Britain, 1865-1930: The Social Construction of Knowledge*. Edinburgh: Edinburgh University Press.

MacKenzie, Donald A (1981b). 'Sociobiologies in Competition: The Biometrician-Mendelian Debate', in Charles Webster (ed.), *Biology, Medicine and Society, 1840-1940*, pp. 243–88. Cambridge: Cambridge University Press.

MacKenzie, J. S. (1896). 'Review: *The Theory of Knowledge: A Contribution to Some Problems of Logic and Metaphysics* by L. T. Hobhouse', *Mind*, 5: 396–410.

McKinnon, Susan (2006). *Neo-Liberal Genetics: The Myths and Moral Tales of Evolutionary Psychology*. Chicago: Chicago University Press.

McLaren, Angus (1978). *Birth Control in Nineteenth-Century England*. London: Croom Helm.

Macleod, Henry Dunning (1872–5). *The Principles of Economical Philosophy*, 2 vols. London: Longmans, Green, Reader, and Dyer.

Macleod, Henry Dunning (1874–5). 'What is Political Economy?', *Contemporary Review*, 25: 871–93.

MacLeod, Roy (2006). 'Consensus, Civility, and Community: Edward Shils, *Minerva*, and the Social Studies of Science'. Paper delivered at the History and Philosophy of Science Informal Seminar, University of Leeds. Publication forthcoming.

MacRae, Donald G. (1972). 'The Basis of Social Cohesion', in W.A. Robson (ed.), *Man and the Social Sciences*, pp. 39–59. London: Allen & Unwin.

Magnello, M. Eileen (1998). 'Karl Pearson's Mathematization of Inheritance: From Ancestral Heredity to Mendelian Genetics (1895–1909)', *Annals of Science*, 55: 35–94.

Mairet, Philip (1957). *Pioneer of Sociology: The Life and Letters of Patrick Geddes*. London: Lund Humphries.

Maloney, John (2005). *The Political Economy of Robert Lowe*. Basingstoke: Palgrave Macmillan.

Martin, A. Patchett (1893). *Life and Letters of the Right Honourable Robert Lowe, Viscount Sherbrooke, GCB, DCI, Etc*, 2 Vols. London: Longmans, Green, and Co.

Matthew, H.C.G. (1973). *The Liberal Imperialists: The Ideas and Policies of a Post-Gladstonian Elite*. London: Oxford University Press.

Mavor, James (1923). *My Windows on the Street of the World*, 2 vols. London: J.M. Dent and Sons Ltd.

Mayr, Ernst (1982). *The Growth of Biological Thought: Diversity, Evolution, and Inheritance*. Cambridge, MA: Belknap Press.

Mazumdar, Pauline M. H. (1992). *Eugenics, Human Genetics and Human Failings: The Eugenics Society, its Sources and its Critics in Britain*. London: Routledge.

—— (1999). 'The Galton Lecture 1998. Eugenics: The Pedigree Years', in Robert Peel (ed.), *Human Pedigree Studies*, pp. 18–44. London: Galton Institute.

Meacham, Standish (1987). *Toynbee Hall and Social Reform, 1880–1914*. New Haven, Conn.: Yale Univ. Press.

Meller, Helen (1990). *Patrick Geddes: Social Evolutionist and City Planner*. London: Routledge.

Metz, Rudolf (1938). *A Hundred Years of British Philosophy*. London: George Allen and Unwin.

Mill, John Stuart (1869). 'Thornton on Labour and Its Claims', *Fortnightly Review*, 5: 505–18; 680–700.

Miller, D. L. (1989). *Lewis Mumford: A Life*. New York: Weidenfeld & Nicholson.

Miller, W. L. (1971). 'Richard Jones: A Case Study in Methodology', *History of Political Economy*, 3: 198–207.

Mills, C. Wright (1959). *The Sociological Imagination*. New York: Oxford University Press.

Mitchell, B. R. (1988). *British Historical Statistics*. Cambridge: Cambridge University Press.

Morgan, Mary (2006). 'Economic Man as Model Man: Ideal Types, Idealization and Caricatures', *Journal of the History of Economic Thought*, 28: 1–27.

Morrell, Jack, and Arnold Thackray (1981). *Gentlemen of Science: Early Years of the British Association for the Advancement of Science*. Oxford: Clarendon Press.

Moseley, H. N. (1882). 'Researches on Animals Containing Chlorophyll', *Nature*, 25: 338.

Mowat, C. L. (1961). *The Charity Organisation Society, 1869–1913: Its Ideas and Work*. London: Methuen.

Mumford, Lewis (1939). *Men Must Act*. Martin Secker and Warburg.

—— (1940). *The Culture of Cities*. Martin Secker and Warburg. (Work originally published in 1938).

—— (1941). *Faith for Living*. Martin Secker and Warburg.

—— (1948). 'Patrick Geddes, Victor Branford, and Applied Sociology in England: the Social Survey, Regionalism, and Urban Planning', in Harry Elmer Barnes (ed.), *An Introduction to the History of Sociology*, pp. 677–95. Chicago: University of Chicago Press.

—— (1955). 'Patrick Geddes', in Lewis Mumford, *The Human Prospect*, H.T. Moore and K.W. Deutsch (eds), pp. 99–114. Boston, MA: Beacon Press. (Article originally published 1950).

—— (1982). *Sketches from Life. The Autobiography of Lewis Mumford: The Early Years*. New York: Dial Press.

Nicholson, Peter P. (1990). *The Political Philosophy of the British Idealists: Selected Studies*. Cambridge: Cambridge University Press.

Nyhart, Lynn K. (1995). 'Natural History and the "New" Biology', in N. Jardine, J. A. Secord, and E. C. Spary (eds), *Cultures of Natural History*, pp. 426–43. Cambridge: Cambridge University Press.

O'Donoghue, Jim, Louise Goulding, and Grahame Allen (2004). 'Consumer Price Inflation Since 1750', *Economic Trends*, 604: 38–46.

Osborne, Thomas, Nikolas Rose, and Mike Savage (2008a). 'Reinscribing the History of British Sociology: Some Critical Reflections', *Sociological Review*, 56: 519–34.

—— (2008b). 'Populating Sociology: Carr-Saunders and the Problem of Population', *Sociological Review*, 56: 552–78.

Parsons, Talcott (1937). *The Structure of Social Action: A Study in Social Theory with Special Reference to a Group of Recent European Writers*. New York: McGraw-Hill Book Company.

Passmore, John (1966). *A Hundred Years of Philosophy*. London: Gerald Duckworth and Co.

Paul, Diane B. (1998). *Controlling Human Heredity: 1865 to the Present*. New York: Humanity.

—— (2003). 'Darwin, Social Darwinism and Eugenics', in Jonathan Hodge and Gregory Radick (eds), *The Cambridge Companion to Darwin*, pp. 214–39. Cambridge: Cambridge University Press.

'P. C. M.' (1889–90). 'The Evolution of Sex', *Nature*, 41: 531–2.

Pearson, Karl (1897–8). 'Mathematical Contributions to the Theory of Evolution. Ancestral Law of Heredity', *Proceedings of the Royal Society of London*, 62: 386–412.

—— (1901). *National Life from the Standpoint of Science: An Address Delivered at Newcastle, November 19th 1900*. London: A.C. & Black.

—— (1914–24). *The Life, Letters and Labours of Francis Galton*, 3 Vols. Cambridge: Cambridge University Press.

Peel, J. D.Y. (1971). *Herbert Spencer: The Evolution of a Sociologist*. London: Heinemann.

Perkins, Harold (2000). ' "Nor all that Glitters...": The Not So Golden Age', in Ian Inkster, Colin Griffin, and Judith Rowbottom (eds), *The Golden Age: Essays in British Social and Economic History, 1850–1870*, pp. 9–26. Aldershot: Ashgate.

Petrilli, Susan (2004). 'Welby , Victoria Alexandrina Maria Louisa, Lady Welby (1837–1912)', in *Oxford Dictionary of National Biography*, online ed., edited by Lawrence Goldman. Oxford: Oxford University Press, http://www.oxforddnb.com/view/article/38619 (accessed 20 June 2011).

Pietarinen, Ahti-Veikko (2009). 'Significs and the Origins of Analytic Philosophy', *Journal of the History of Ideas*, 70: 467–90.

Platt, Jennifer (2003). *The British Sociological Association: A Sociological History*. Durham: Sociology Press.

Political Economy Club of London (1876). *Revised Report of the Proceedings at the Dinner of 31st May, 1876, Held in Celebration of the Hundredth Year of the Publication of the 'Wealth of Nations'. Right Hon. W. E. Gladstone in the Chair*. London: Longmans, Green, Reader & Dyer.

Political Economy Club of London (1882). *Minutes of Proceedings, 1821-1882, Roll of Members, and Questions Discussed. Volume IV.* London: Political Economy Club of London.

Porter, Theodore M. (1994) 'Rigor and Practicality: Rival Ideals of Quantification in Nineteenth Century Economics', in Philip Mirowski (ed.), *Natural Images in Economic Thought: "Markets Read in Tooth and Claw,"* pp. 128–70. New York: Cambridge University Press.

—— (2004). *Karl Pearson: The Scientific Life in a Statistical Age.* Princeton: Princeton University Press.

—— (2011). 'Reforming Vision: The Engineer Le Play Learns How to Observe Society Sagely', in Lorraine Daston and Elizabeth Lunbeck (eds), *Histories of Scientific Observation*, pp. 281–302. Chicago: University of Chicago Press.

Price, Bonamy (1879). 'Address on Economy and Trade', *Transactions of the National Association for the Promotion of Social Science: Cheltenham Meeting, 1878*: 116–34.

'Proceedings of the Forty-Third Anniversary Meeting' (1877). *Journal of the Statistical Society of London*, 40: 342–6.

Proctor, Robert (1991). *Value-Free Science? Purity and Power in Modern Knowledge.* Cambridge, MA: Harvard University Press.

Provine, William B. (1971). *The Origins of Theoretical Population Genetics.* Chicago: University of Chicago Press.

Radick, Gregory (2007). *The Simian Tongue: The Long Debate About Animal Language.* Chicago: University of Chicago Press.

—— (2008). 'Introduction: Why What if?', *Isis*, 99: 547–51.

Radick, Gregory and Graeme Gooday (2004). 'Patrick Geddes (1854–1932)', in Bernard Lightman (ed.), *The Dictionary of Nineteenth-Century British Scientists*, 4 vols., Vol. 2, pp. 764–8. Bristol: Thoemmes.

Radkau, Joachim (2009). *Max Weber: A Biography*, trans. by Patrick Camiller. Cambridge: Polity.

Reid, G. Archdall (1906). 'The Biological Foundations of Sociology', *Sociological Papers*, 3: 3–27.

Renwick, Chris (2009). 'The Practice of Spencerian Science: Patrick Geddes' Biosocial Programme, 1876–89', *Isis*, 100: 36–57.

—— (2011). 'From Political Economy to Sociology: Francis Galton and the Social-Scientific Origins of Eugenics', *The British Journal for the History of Science*, 44: forthcoming.

—— (2011). "Observation and Detachment: William Beveridge and the Natural Bases of Social Science," unpublished manuscript and paper presented to the *History of Political Economy* conference "A History of Observation in Economics," Duke University, 18th April 2011.

Renwick, Chris and Richard C. Gunn (2008). 'Demythologizing the Machine: Patrick Geddes, Lewis Mumford, and Classical Sociological Theory', *Journal of the History of the Behavioral Sciences*, 44: 59–76.

Richards, Robert J. (1987). *Darwinism and the Emergence of Evolutionary Theories of Mind and Behaviour.* Chicago: University of Chicago Press.

—— (2004). 'If this be Heresy: Haeckel's Conversion to Darwinism', in Abigail Lustig, Robert J. Richards, and Michael Ruse (eds), *Darwinian Heresies*, pp. 101–30. Cambridge: Cambridge University Press.

Richardson, Angelique (2003). *Love and Eugenics in the Late Nineteenth Century: Rational Reproduction and the New Woman*. Oxford: Oxford University Press.

Ringer, Fritz (2004). *Max Weber: An Intellectual Biography*. Chicago: University of Chicago Press.

Ritchie, David G. (1901). *Darwinism and Politics*. London: Swan Sonnenschein and Co.

Rogers, James E. Thorold (1888). *The Economic Interpretation of History. Lectures Delivered in Worcester College Hall, Oxford, 1887-8*. London: T. Fisher Unwin.

Rose, Hilary and Steven Rose (eds) (2001). *Alas Poor Darwin: Arguments Against Evolutionary Psychology*. London: Vintage.

Rouse, Joseph (2010). 'Why Write Histories of Science?', *History of the Human Sciences*, 23:100–4.

Royle, Edward (1980). *Radicals, Secularists and Republicans: Popular Freethought in Britain, 1866-1915*. Manchester: Manchester University Press.

Runciman, W. G. (1998). *The Social Animal*. London: Harper Collins.

—— (2005). 'Introduction', in A. H. Halsey and W. G. Runciman (eds), *British Sociology Seen from Without and Within*, pp. 1–9. Oxford: Oxford University Press for the British Academy.

—— (2008). 'Forgetting the Founders', *Sociological Review*, 56: 358–69.

Ruse, Michael (1979). *Sociobiology: Sense or Nonsense?* Dordrecht: D. Reidel Publishing Company.

—— (1996). *Monad to Man: The Concept of Progress in Evolutionary Biology*. Cambridge, MA: Harvard University Press.

—— (2004). 'Adaptive Landscapes and Dynamic Equilibrium: The Spencerian Contribution to Twentieth-Century American Biology', in Abigail Lustig, Robert J. Richards, and Michael Ruse (eds), *Darwinian Heresies*, pp. 131–50. Cambridge: Cambridge University Press.

Russell, Bertrand (1907). 'The Development of Morals', *Independent Review*, 12: 204–10.

—— (1914). *Our Knowledge of the External World*. London: Allen and Unwin.

Sapp, Jan (1994). *Evolution by Association: A History of Symbiosis*. New York: Oxford University Press.

Savage, Mike (2008). 'Elizabeth Bott and the Formation of Modern British Sociology', *Sociological Review*, 56: 579–605.

Schabas, Margaret (1997). 'Victorian Economics and the Science of Mind', in Bernard Lightman (ed.), *Victorian Science in Context*, pp. 72–93. Chicago: University of Chicago Press.

—— (2003). 'British Economic Theory from Locke to Marshall', in Theodore M. Porter and Dorothy Ross (eds), *The Cambridge History of Science Volume 7: The Modern Social Sciences*, pp. 171–82. New York: Cambridge University Press.

—— (2005). *The Natural Origins of Economics*. Chicago: University of Chicago Press.

Schaffer, Gavin (2008). *Racial Science and British Society, 1930-1962*. Basingstoke: Palgrave Macmillan.

Schmaus, Warren (2004). *Rethinking Durkheim and His Tradition*. Cambridge: Cambridge University Press.

Schumpeter, Joseph (1976). *History of Economic Analysis*. Elizabeth Boody Schumpeter (ed.). New York: Oxford University Press.

Scott, C. P. (1931). 'Introduction', in J. A. Hobson and Morris Ginsberg, *L. T. Hobhouse: His Life and Work*, pp. 7–9. London: George Allen and Unwin.

Scott, John, and Christopher T. Husbands (2007). 'Victor Branford and the Building of British Sociology', *Sociological Review*, 55: 460–84.

Searle, G. R. (1990). *The Quest for National Efficiency: A Study in British Politics and Political Thought, 1899-1914*. London: Ashfield.

Searle, G. R (1998). *Morality and the Market in Victorian Britain*. Oxford: Clarendon Press.

Secord, James A. (2004). 'Knowledge in Transit', *Isis*, 95: 654–72.

Segerstråle, Ullica (2000). *Defenders of the Truth: The Sociobiology Debate*. Oxford: Oxford University Press.

Semmel, Bernard (1960). *Imperialism and Social Reform: English Social-Imperial Thought, 1895-1914*. London: George Allen and Unwin.

Seth, James (1908). 'Review: *Morals in Evolution: A Study in Comparative Ethics* by L. T. Hobhouse,' *International Journal of Ethics*, 18: 375–81.

Shapin, Steven (2009). 'Review of *New Dictionary of Scientific Biography*', Noretta Koertge (ed.), *British Journal for the History of Science*, 42: 116–17.

—— (2011). 'Good Housekeeping: Review of *William Petty and the Ambitions of Political Arithmetic* by Ted McCormick', *London Review of Books*, 33: 19–21.

Shaw, George Bernard (ed.) (1900). *Fabianism and Empire*. London: Grant Richards.

Shils, Edward (1980). *The Calling of Sociology and Other Essays in the Pursuit of Learning*. Chicago: Chicago University Press.

—— (1985). 'On the Eve: A Prospect in Retrospect', in Martin Bulmer (ed.), *Essays on the History of British Sociological Research*, pp. 165–78. Cambridge: Cambridge University Press. (Article originally published 1960).

Sidgwick, Henry (1887). *The Principles of Political Economy*, 2nd edn. London: Macmillan & Co.

Silver, Catherine Bodard (1982). 'Introduction', in Frédéric Le Play, *Frédéric Le Play on Family, Work, and Social Change*, ed., trans., and introduction by Catherine Bodard Silver, pp. 1–134. Chicago: University of Chicago Press.

Smith, P. J. (1980). 'Planning as Environmental Improvement: Slum Clearance in Victorian Edinburgh', in Anthony Sutcliffe (ed.), *The Rise of Modern Planning, 1800-1914*, pp. 99–133. London: Mansell.

Smith, Roger (1997). *The Fontana History of the Human Sciences*. London: Fontana.

Sociological Society (1906). *Second Annual Report with Rules and List of Members*. London: Sociological Society.

'Sociology in England' (1904). *Speaker*, 10: 65.

Soloway, Richard A. (1990). *Demography and Degeneration: Eugenics and the Declining Birthrate in Twentieth-Century Britain*. Chapel Hill: University of North Carolina Press.

Spencer, Herbert (1851). *Social Statics, or the Essential Conditions to Human Happiness Specified, and the First of them Developed*. London: Williams and Norgate.

—— (1858) 'The Development Hypothesis', in Herbert Spencer, *Essays: Scientific, Political and Speculative*, pp. 389–95. London: Longman & Roberts. (1st published 1852)

—— (1864-7). *The Principles of Biology*, 2 Vols. London: Williams and Norgate.

—— (1867). *First Principles*, 2nd edn. London: Williams and Norgate. (1st edn published 1862).

—— (1868–74). *Essays: Scientific, Political, and Speculative*, 3 Vols. London: Williams and Norgate.

—— (1870–2). *The Principles of Psychology*, 2nd edn, 2 Vols. London: Williams & Norgate.

—— (1873). *The Study of Sociology*. London: H.S. King.

—— (1879). *The Data of Ethics*. London: Williams and Norgate.

—— (1887). *The Factors of Organic Evolution*. London: Williams and Norgate.

—— (1896). *Principles of Sociology*, 3 Vols. London: Williams and Norgate.

Staley, Thomas W. (2009a). 'The Journal *Mind* in its Early Years, 1876–1920: An Introduction', *Journal of the History of Ideas*, 70: 259–63.

—— (2009b). 'Keeping Philosophy in Mind: Shadworth H. Hodgson's Articulation of the Boundaries of Philosophy and Science', *Journal of the History of Ideas*, 70: 289–315.

Stalley, Marshall (ed.) (1972). *Patrick Geddes: Spokesman for Man and the Environment*. New Brunswick: Rutgers University Press.

Stephen, Walter (ed.) (2004). *Think Global, Act Local: The Life and Legacy of Patrick Geddes*. Edinburgh: Luath Press.

Stigler, George J. (1965). 'Statistical Studies in the History of Economic Thought', in George J. Stigler, *Essays in the History of Economic Thought*, pp. 31–50. Chicago: University of Chicago Press. (Originally published in 1964.)

Stigler, Stephen M. (1986). *The History of Statistics: The Measurement of Uncertainty Before 1900*. Cambridge, MA: Harvard University Press.

Stocking Jnr., George W. (2001). 'Dr. Durkheim and Mr. Brown: Comparative Sociology at Cambridge in 1910. Edited by George W. Stocking, Jnr.', in George W. Stocking Jnr. (ed.), *Delimiting Anthropology: Occasional Essays and Reflections*, pp. 106–30. Madison: University of Wisconsin Press.

Studholme, Maggie (1997). *British Sociology and the Issue of the Environment*. Unpublished PhD Thesis, University of Bristol.

—— (2007). 'Patrick Geddes: Founder of Environmental Sociology', *Sociological Review*, 55: 441–59.

—— (2008). 'Patrick Geddes and the History of Environmental Sociology in Britain: A Cautionary Tale', *Journal of Classical Sociology*, 8: 367–91.

Studholme, Maggie, John Scott, and Christopher T. Husbands (2007). 'Doppelgängers and Racists: On Inhabiting Alternative Universes. A Reply to Steve Fuller's "A Path Better Not to Have Been Taken"', *Sociological Review*, 55: 816–22.

Sweeney, Gerald (2001). *'Fighting for the Good Cause': Reflections on Francis Galton's Legacy to American Hereditarian Psychology*. Philadelphia: American Philosophical Society.

'The Sociological Society—Its Origins and Aims' (1904). *Sociological Papers*, 1: 284–6.Thomson, J. Arthur (1906).

'The Sociological Appeal to Biology', *Sociological Papers*, 3: 157–85.

Todhunter, I. (1876). *William Whewell, D. D, Master of Trinity College, Cambridge. An Account of His Writings With Selections from his Literary Correspondence* 2 Vols. London: Macmillan & Co.

Tönnies, Ferdinand (1899). 'Philosophical Terminology', *Mind*, 8: 289–332; 467–91.
—— (2001). *Community and Civil Society*, (ed.) José Harris, trans. José Harris and Margaret Hollis. Cambridge: Cambridge University Press. (1st published 1887).
Topham, Jonathan R. (2004). 'A View from the Industrial Age', *Isis*, 95: 431–42.
Tribe, Keith (2011). 'Founders of the Political Economy Club (*act.* 1821–1829)', in *Oxford Dictionary of National Biography*, online ed., edited by Lawrence Goldman. Oxford: Oxford University Press. http://0-www.oxforddnb.com. wam.leeds.ac.uk/view/theme/95369 (accessed 20 June 2011.
Vincent, Andrew, and Raymond Plant (1984). *Philosophy, Politics and Citizenship: The Life and Thought of British Idealists*. Oxford: Blackwell.
Waller, John C. (2001). 'Gentlemanly Men of Science: Sir Francis Galton and the Professionalization of the British Life-Sciences', *Journal of the History of Biology*, 34: 83–114.
Webb, Beatrice (1906). 'Methods of Investigation', *Sociological Papers*, 3: 345–51.
—— (1926). *My Apprenticeship*. London: Longmans, Green and Co.
Webb, Sidney and Beatrice Webb (1897). *Industrial Democracy*, 2 Vols. London: Longmans, Green and Co.
—— (1906–29). *English Local Government*, 9 Vols. London: Longmans, Green.
Weber, Max (1949). *The Methodology of the Social Sciences*, Edward Shils(ed.). New York: Free Press.
—— (1958). *The Protestant Work Ethic and the Spirit of Capitalism*, trans. Talcott Parsons, foreword by R.H. Tawney. New York: Charles Scribner's and Sons. (1st published in two parts in 1904–05).
Welby, Victoria Lady (1903). *What is Meaning? Studies in the Development of Significance*. London: Macmillan & Co.
—— (1911). *Significs and Language. The Articulate Form of Our Expressive and Interpretative Resources*. London: Macmillan & Co.
—— (1929). *Echoes of Larger Life: A Selection from the Earlier Correspondence of Victoria Lady Welby*, Mrs. Henry Cust (ed.). London: Jonathan Cape.
—— (1931). *Other Dimensions: A Selection from the Later Correspondence of Victoria Lady Welby*, Mrs. Henry Cust (ed.). London: Jonathan Cape.
Weldon, W. F. R. (1893). 'On Certain Correlated Variations' in *Carcinus Moenas*', *Proceedings of the Royal Society of London*, 54: 318–33.
—— (1894–5). 'Remarks on Variation in Animals and Plants. To Accompany the First Report of the Committee for Conducting Statistical Inquiries into Measurable Characteristics of Plants and Animals', *Proceedings of the Royal Society of London*, 57: 379–82.
Wells, H. G. (1906). 'The So-Called Science of Sociology', *Sociological Papers*, 3: 357–69.
Welter, Volker (2002). *Biopolis: Patrick Geddes and the City of Life*. Cambridge, MA: MIT Press.
Westermarck, Edward (1904). 'On the Position of Woman in Early Civilisation', *Sociological Papers*, 1: 145–60.
Wilson, E. O. (1975). *Sociobiology: The New Synthesis*. Cambridge, MA: Harvard University Press.
Winter, James (1976). *Robert Lowe*. Toronto: University of Toronto Press.

Yeo, Eileen Janes (1996). *The Contest for Social Science: Relations and Representations of Gender and Class.* London: Rivers Oram Press.

Young, Cristobal (2009). 'The Emergence of Sociology from Political Economy in the United States: 1890-1940', *Journal of the History of the Behavioral Sciences*, 45: 91–116.

Young, Robert M. (1970). *Mind, Brain and Adaptation in the Nineteenth Century: Cerebral Localization and its Biological Context from Gall to Ferrier.* Oxford: Clarendon Press.

Zueblin, Charles (1899). 'The World's First Sociological Laboratory', *American Journal of Sociology*, 4: 577–92.

Index